Generis
PUBLISHING

Autotransformer discrete alternating voltage regulators

Analyses, simulations, experiments

Emil Ivanov Panov
Emil Stefanov Barudov
Milena Dimitrova Ivanova

Title: Autotransformer discrete alternating voltage regulators

Analyses, simulations, experiments

ISBN: 979-8-88676-946-3

Author: Emil Ivanov Panov, Emil Stefanov Barudov,Milena Dimitrova Ivanova

Cover image: www.pixabay.com

Publisher: Generis Publishing
Online orders: www.generis-publishing.com
Contact email: info@generis-publishing.com

The presented book is a collection of selected papers that the author's team has been developing for more than 20 years in the field of AC step regulators. The number of publications on the proposed topic comprises more than 50 works, only 14 of which are included in the book. The research done has been published in numerous books, journals, annuals and international conferences.

In addition to the authors mentioned, the research group over the years included the group leader Prof. D.Sc. Eng. Stefan Todorov Barudov, who was the leading figure in the research team, as well as some other researchers with episodic participation.

With this collection of papers, the research team hopes to be useful to other researchers, developers and engineers in their work on the development, computer modelling and operation of step voltage regulators.

CV: Stefan Todorov Barudov graduated from the Technical University of Varna in major “Radio Engineering”, MSc degree, in 1971 and acquired the qualification "Engineer". In 1983, he obtained PhD degree at the Technical University of Varna. He had been rector of the Technical University of Varna from 1999 to 2007 and a professor at the same university from 2007 to 2015. Other memberships: IEEE member, president of the Black Sea Universities Network (BSUN), honorary member of the Union of Electronics, Electrical Engineering and Communications. Honours: 2002 - University of Abertay Dundee, Scotland – „Doctor Honoris Causa”, 2006 – Member of the International Academy of Informatics. For more than 30 years he was responsible for the development, the design and the implementation of the autotransformer discrete alternating voltage regulators (ADAVR) as a scientific direction.

Fields of scientific research: Processes and devices for control of electrical discharges, electronics, power electronics, laser technology

e-mail: sbarudov@abv.bg

CV: Emil Ivanov Panov graduated from the Technical University of Varna in major “Communications”, MSc degree, in 1984 and acquired the qualification "Engineer". In 1985, he became assistant-professor at the Department of Theoretical Electrical Engineering and Instrumentation in the same university. In 1988, he received second MSc degree in major “Scientific and Technical Development” at the Academy of Social Sciences and Social Management - Varna. In 1994, he defended a PhD thesis on the topic “Exploring the Possibilities of Applying the Taylor Transform Method for the Analysis of Switched-Capacitor Circuits" and obtained PhD degree at the Technical University of Varna. Since 1999, he is associate professor. He had been head of the department "Theoretical Electrical Engineering and Instrumentation" from 1999 till 2007 and then, from 2013 till 2020 he was a vice dean on "Scientific and Human Resources Development" of the Electrical Faculty of the Technical University of Varna. In March 2023, he retired from the Technical University of Varna. He is a member of the Union of the Scientists in Bulgaria since 1985. For more than 20 years he was responsible for the analysis, the numeric algorithms, the computer simulations and the models in the research of the autotransformer discrete alternating voltage regulators as a scientific direction. Honours: 1984 – 2 gold medals for excellent success and active community activity.

Fields of scientific research: Analysis of switched-capacitor circuits, Taylor transform method, analysis and synthesis of electronic circuits, analysis of power electronic circuits, analysis of non-linear circuits, computer simulations of electric circuits, numeric algorithms for computer simulations of the processes in electric circuits, author of a modified variational method for analysis of electric circuits, special relativistic theory, he has developed the relativistic electric circuits analysis, author of the Rotary theory of the electromagnetic field in classic and relativistic form

e-mail: eipanov@yahoo.com

CV: Emil Stefanov Barudov graduated from the Technical University of Varna in major "Electrical Power Supply and Electrical Equipment of Water Transport", MSc degree, in 1998 and acquired the qualification " Master - Engineer ". In 2014, he defended a PhD thesis on the topic "Research and Analysis of Electrical Processes in Circuits with Devices for Discrete Control of the Magnitude of Alternating Voltage". Since 2015, he has been an associate professor at the Department of Electric Engineering at the Engineering Faculty of the "Nikola Vaptsarov" Naval Academy - Varna. Honours: 2004 – Gold medal at the LX^{-th} International Technical Fair Plovdiv – "Gamma Single-Phase Oil Step-Down Transformers for Pole Mounting Type 2TMXX 720/0,23 for 20kV Set with Regulators" (participating in a team).

Fields of scientific research: Power electronics, semiconductor converters, high-voltage engineering, electromagnetic compatibility, electrical safety

e-mail: ugl@abv.bg

CV: Milena Dimitrova Ivanova graduated from the Technical University of Varna in major "Communication Engineering and Technologies" BSc degree (in 2005) and MSc degree (in 2007) at the Technical University - Varna and acquired the qualification "Master - Engineer". In 2016, she successfully defended her PhD thesis on the topic "Electrical Processes in Circuits for Generating a High-Voltage Discharge Pulse in a Liquid Medium" in the scientific major "Electrical Power Supply and Electrical Equipment in Industry (of the Ship)" at the "Nikola Vaptsarov" Naval Academy – Varna. Since 2005, she has been a part-time assistant, and since 2007, a full-time assistant at the Department of Electric Power Engineering at the Technical University of Varna. Since 07.2023 she is an associate professor at the department.

Fields of scientific research: Electrical engineering materials, high voltage discharges in liquid, technical safety, earthing grids, diagnostics of cable power lines

e-mail: sunny.mim@gmail.com

CONTENTS

Chapter 1. Parameters, Basic Equations, and Operation of the ADAVR 9
Chapter 2. Analysis, Algorithms, Computer Programs, Simulations, Experiments 17
Chapter 3. Comparison Between the Approximate and Precise Models of the ADAVR 74
Chapter 4. Exploration on the Commutation Regimes of the ADAVR 84
Chapter 5. Analysis of the Sensitivity of ADAVR 102
Chapter 6. Errors Between the Results of the Simulations and Physical Experiments. 113
Chapter 7. Results of the Experiments of ADAVR with R, RL and RC Loads 122
Chapter 8. Exploration on the Efficiency of the ADAVR 130
Chapter 9. Protection Regimes in the ADAVR 155
Chapter 10. Two-Port Parameters of the ADAVR 165
Reference of Publications on the Topic 175

Chapter 1. Parameters, Basic Equations, and Operation of the ADAVR

SWITCHING PROCESSES IN A STEP VOLTAGE REGULATOR

BARUDOV Emil, BARUDOV Stefan, PANOV Emil
(Published in "Acta Universitatis Pontica Euxinus", Vol. 4, N. 1, 2005, pp. 21÷25, ISSN 1312-1669)
Technical University – Varna, 1 Sudentska str., 9010 Varna
e-mail: ugl@abv.bg

Abstract: The voltage of the separate consumers in limited power networks can be maintained within the allowed boundaries by automated step change-over of an autotransformer. The transient phenomenon of a thyristor switch breakeover at the boost winding of the autotransformer is studied.

Keywords: step changeover, auto-transformer, regulator, model, transient process.

Introduction: There are electrical networks where the phase voltage varies far above the allowable limits depending on the operational modes of some of the ultimate consumers. This forces application of individual regulation of the phase voltage directly at the very consumers. The regulation can be effected by step-wise automated changeover of the autotransformer terminals.

The paper considers the transient process of thyristor switches changeover at the boost stage. On the basis of a mathematical model describing the voltage regulator as well the switching processes, a MATLAB AVTO1 program was developed, which makes visual indication of the transient process and calculates some parameters of the real model. The program evaluates the magnitude of the current during switching and the length of the transient, which allows efficient and economically feasible calculation of the regulator parameters.

In order to define initially the parameters necessary to elaborate the model, we carried out no load and short circuit experiments with a 4kVA single-phase autotransformer and the data were used to define the coefficients and to construct a parametric matrix. The transient process should be considered in three intervals:

- First interval – the voltage slowly drops from 170V to 165V; K_3 is open, K_2 is closed, K_1 and K_0 are open (the last but one level of the voltage regulator is switched on).
- Second interval – K_3 closes (with phase delay from e(t) zero level equal to 90°), but K_2 is not opened – an interturn short circuit takes place.
- Third interval – the switch K_3 is closed and K_2 opens when the current $i_4(t)$ equals zero.

Fig. 1 shows the voltage regulator model with the existing magnetic links among the autotransformer sections. The received values of the parameters are presented in Table 2. The method of conditional linearization is used for their definition. The description of the electrical equilibrium of the voltage regulator is made with the application of mesh analysis as it provides sets of differential equations with minimal dimensionality.

The complex method is used to solve the differential equations in the first interval (slow decrease of regulator input voltage from 170V down to 165V. In the second and the third time intervals after the switching, the analysis is conducted in the time domain, while the systems of differential equations are reduced to algebraic systems by the Euler method and are solved by Gauss – Jordan elimination. The graphs of the processes clearly display lack of time errors.

Figures 2, 3 and 4 show the equivalent circuits of the examined regulator within three time intervals. Table 1 presents the correlation among the currents of the loops. The study was conducted for active loads (a dominant case in households – most of the consumers are heating elements and domestic electronics) with the assumption that K_0 – K_3 are ideal switches, which gives sufficiently accurate notion about the behavior of the processes. The first interval is studied in the complex – frequency domain, while the second and the third intervals are studied in the time domain.

The sets of differential equations (1), (2) and (3) describe the electrical and magnetic processes in the studied time intervals. The equations are solved by the program AVTO1 using MATLAB. The first system (1) is solved by the Gauss-Jordan method in the complex – frequency domain. The second and the third systems (2) and (3) are solved by the Euler method for numerical solution of a set of differential equations.

Figures 5, 6, 7, 8 and 9 display the transient processes at changeover of the switches for i_1, i_2, i_3, i_4 and i_0 currents, respectively. The graphs clearly show the three intervals of the process – initial steady harmonic current, transient process and flow of a new steady harmonic current (with changed value, accordingly).

There is no second transient process in the beginning of the third interval, due to the accurate switching off of K_2.

The computer simulation of the studied switchings in the voltage regulator ascertained about 12,5mS duration of the transient process. This duration corresponds to the emergency state of operation of the autotransformer in case of a sequential switching to regulate the output voltage.

CONCLUSION

The suggested model of electronic voltage regulator allows carrying out computer analysis of the transient processes due to changeover of the thyristor switches.

The analyses were compared with experimental data obtained from the physical examination of the electronic regulator. The data of the both examinations agree with sufficient accuracy and are an index for the adequacy of the established model.

The results of the study can be used successfully to design similar electronic regulators and to calculate correctly their thyristor switches, with the opportunity to trace in real time their performance and modes of operation.

REFERENCES

[1] Minchev M., Penchev P., Contactless Aparatuses, Tehnika, 1976. (in Bulgarian)

[2] Porudominskiy V. Switching De-vices for Transformers under Load, Energia, 1974. (in Russian)

[3] Lipkowskiy K., Transformer-Switching Actuator Structures of the Alternating Current Converters, Naukova Dumka, 1983. (in Russian)

[4] Barudov St., Dimitrov D., Barudov E. Functional Performance of Transformers in a Noncontact Voltage Stabilizer, SIELA 2001. (in Bulgarian)

1. Steady state mode at $\dot{E} = 165V$ (first interval).

$$(1)\quad \begin{cases} (r_1 + j\omega L_1).\dot{I}_1' - [r_1 + j\omega.(L_1 + M_{123})].\dot{I}_2' = \dot{E} \\ -[r_1 + j\omega.(L_1 + M_{123})].\dot{I}_1' + [(R_T + r_1 + r_2 + r_3) + j\omega.(L_1 + 2M_{123} + L_{23})].\dot{I}_2' = 0 \end{cases}$$

2. First transient process after the switching of K_3 (second interval)

$$(2)\quad \begin{cases} L_1.\dfrac{di_1'(t)}{dt} - (L_1 + M_{123}).\dfrac{di_2'(t)}{dt} - M_{14}\dfrac{di_3'(t)}{dt} = -r_1.i_1'(t) + r_1.i_2'(t) + e_m.\sin(\omega t + \phi) \\ -(L_1 + M_{123}).\dfrac{di_1'(t)}{dt} + (L_1 + 2M_{123} + L_{23}).\dfrac{di_2'(t)}{dt} + (M_{14} + M_{234}).\dfrac{di_3'(t)}{dt} = r_1.i_1'(t) - (R_T + r_1 + r_2 + r_3).i_2'(t) \\ -M_{14}.\dfrac{di_1'(t)}{dt} + (M_{14} + M_{234}).\dfrac{di_2'(t)}{dt} + L_4.\dfrac{di_3'(t)}{dt} = -r_4.i_3'(t) \end{cases}$$

3. Third interval - after K_2 switching

$$(3)\quad \begin{cases} L_1.\dfrac{di_1(t)}{dt} - (L_1 + M_{1234}).\dfrac{di_2(t)}{dt} = -r_1.i_1(t) + r_1.i_2(t) + e_m.\sin(\omega t + \phi) \\ -(L_1 + M_{1234}).\dfrac{di_1(t)}{dt} + (L_1 + 2M_{1234} + L_{234}).\dfrac{di_2(t)}{dt} = r_1.i_1(t) - (R_T + r_1 + r_2 + r_3 + r_4).i_2(t) \end{cases}$$

Table 1

First Interval	Second Interval	Third Interval
$i_3(t) = 0$	$i_4(t) = i_2(t) - i_3(t)$	$i_2(t) = i_3(t)$
$i_2(t) = i_4(t)$	$i_0(t) = i_1(t) - i_2(t)$	$i_4(t) = 0A$
$i_0(t) = i_1(t) - i_2(t), i_t(t) = i_2(t)$	$i_t(t) = i_2(t)$	$i_0(t) = i_1(t) - i_2(t), i_t(t) = i_2(t)$

Table 2

For U_{in} = 165V	R_T=3,3846 Ω	M_{12}=0,281466H
r_1=0,106Ω	w_1=155 turns	$M_{1234}=M_{12}+M_{13}+M_{14}$=1,0883H
$r_2=3,5.10^{-3}\Omega$	w_2=15 turns	M_{13}=0,3377592H
$r_3=4,1.10^{-3}\Omega$	w_3=18 turns	M_{14}=0,46911H
$r_4=5,7.10^{-3}\Omega$	w_4=25 turns	M_{23}=32,6863mH
L_1=2,9104H		M_{24}=45,3975mH
$L_2=27,2566.10^{-3}H$		$M_{123}=M_{12}+M_{13}$=0,61923H
$L_3=39,2495.10^{-3}H$		$M_{234}=M_{24}+M_{34}$=0,10003H
$L_4=75,78.10^{-3}H$		$L_{234}=L_2+L_3+L_4+2M_{23}+2M_{34}+2M_{24}$= =0,40788H
		$L_{23}=L_2+L_3+2M_{23}$=0,13204H

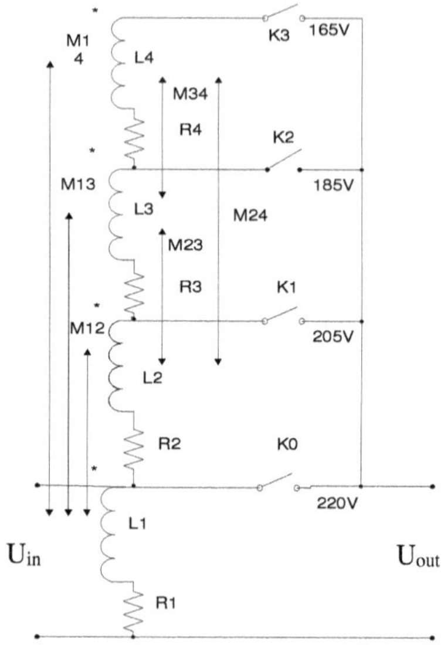

Fig. 1. Equivalent circuit of the step voltage regulator.

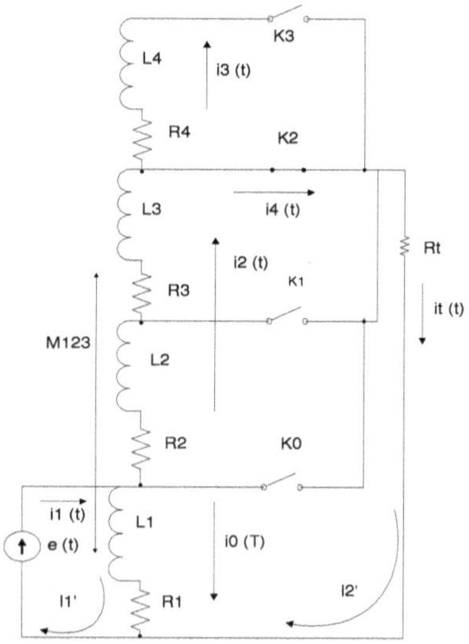

Fig. 2. Equivalent circuit for the first interval during the steady state mode.

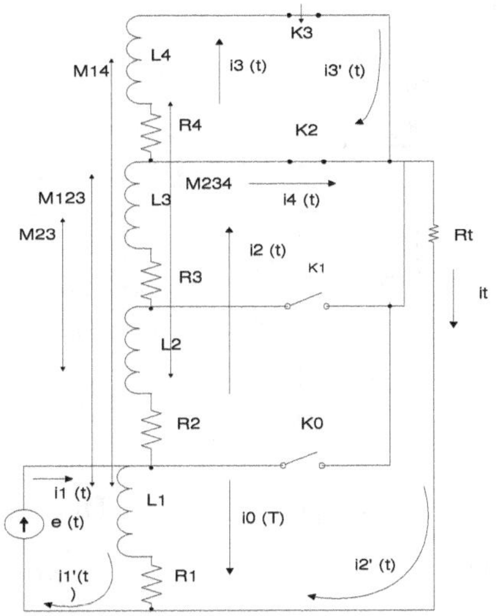

Fig. 3. Equivalent circuit for the second interval during the transient process.

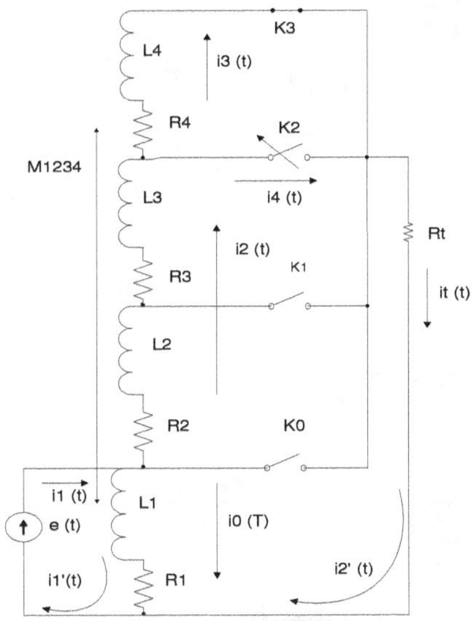

Fig. 4. Equivalent circuit for the third interval during the new steady state mode.

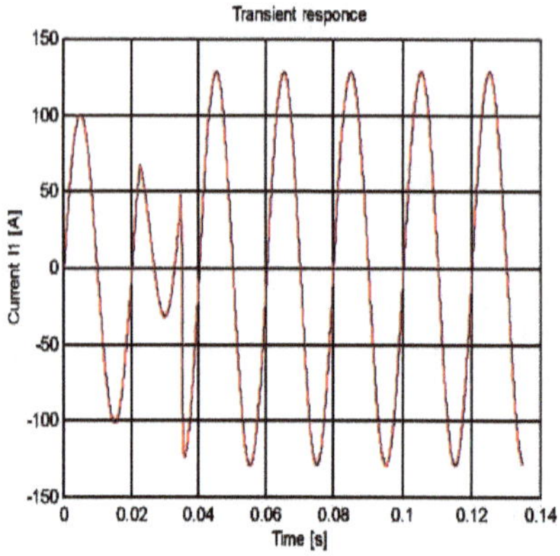

Fig. 5. Diagram of the process for the electric current i_1.

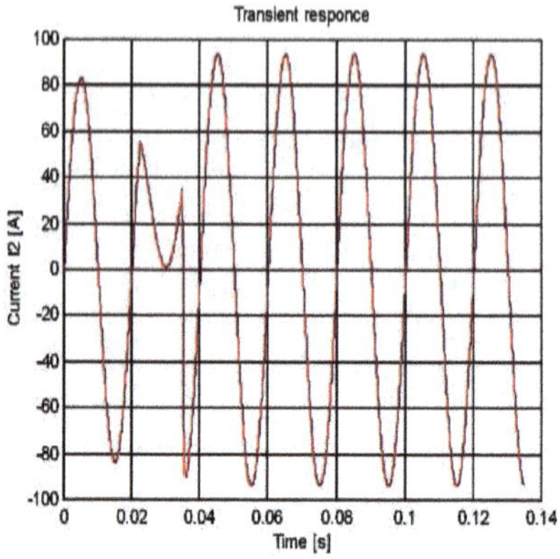

Fig. 6. Diagram of the process for the electric current i_2.

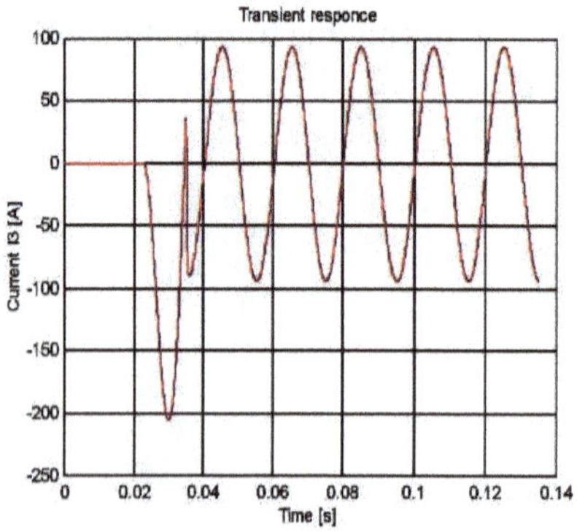

Fig. 7. Diagram of the process for the electric current i_3.

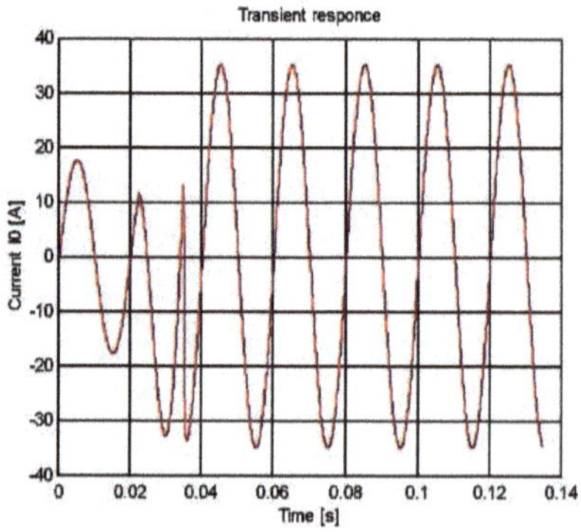

Fig. 8. Diagram of the process for the electric current i_0.

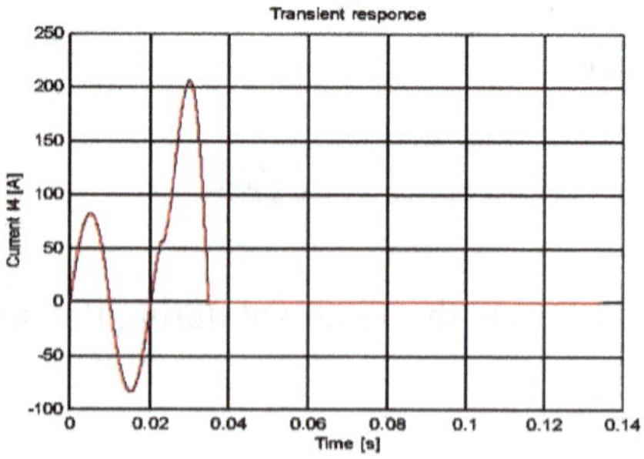

Fig. 9. Diagram of the process for the electric current i_4.

Chapter 2. Analysis, Algorithms, Computer Programs, Simulations, Experiments

ANALYSIS OF ELECTRICAL PROCESSES IN ALTERNATING VOLTAGE CONTROL SYSTEMS

E. Panov, E. Barudov, S. Barudov

(Published on 12-th International Symposium "Materials, Methods&Technologies (MMT), 11-15 June 2010, Sunny Beach, Bulgaria, published in Journal of International Scientific Publications: Materials, Methods & Technologies, Volume 4, Part 1, 2010, pp. 154 – 182, ISSN: 1313 2539, http://www.science-journals.eu.)

Technical University-Varna, 1 "Studentska" str., 9010-Varna, Bulgaria

Abstract

One of the problems concerning the quality of electric power refers to amplitude control of supply voltage alternating in time. This paper presents an analysis of the problems of autotransformer discrete alternating voltage regulators (ADAVR) through simulation models while accounting for the specific features of a particular circuit design. These models are based on the non-linear parameters of the constructive components, such as: windings and ferromagnetic core in autotransformers, thyristor commutators, switch-off assemblies and load. The analysis has been basically made by the state variables method in the temporal field. A special algorithm has been developed for part-by-part solving of the obtained sparse matrix equations of cell structure, which have almost singular matrices. The obtained solutions show insignificant error accumulation from the computations and great stability in time. An automated computer program, AVTO, has also been developed to simulate ADAVR processes in MATLAB environment. The program operates with parameters of autotransformer windings, parameters of ferromagnetic core, parameters of thyristor switches and these of switch-off assemblies. The program analyzes all three intervals inherent to the commutation process and represents in series the behaviour of the individual quantities in time. The data from the simulated processes have been experimentally verified using a prototype and have shown a good correspondence. A system version for virtual design of discrete voltage regulators has been put forward.

Keywords: *autotransformer discrete alternating voltage regulator, semiconductor commuting elements, thyristor, commutation process*

1. INTRODUCTION

Recently, there has been an increasing interest in a more precise parameter control of the electric power supplied to industries and homes [1]. An element of electric power quality is the range of alteration of the input supply voltage amplitude. The supply voltage control can be implemented through autotransformer discrete alternating voltage regulators (ADAVR) with semiconductor commuting elements (SCE) [2, 3]. Design and construction of ADAVR with SCE require modern approaches, such as computer analysis and simulation of the complex processes occurring in devices of this class [1, 2, 3, 4, 5]. The major problem here is the development of adequate, precise models and algorithms for analysis which should enable the achievement of accurate solutions and true simulation of ADAVR responses in a variety of operative modes and in emergent situations as well [6, 7, 8]. The purpose of all this is economy of financial means and time during engineering and manufacturing of such products.

2. ANALYSIS

The amplitude range control of input supply voltage can be achieved through automatic switch-over of the transformer terminals (autotransformer). The terminals can be located on the side of the mains or on the side of the load [1]. Fig. 1 shows a circuit equivalent to one of the commonly used ADAVR with four SCE. The circuit includes the magnetic circuit parameters, commuting groups, switch-off assemblies, parameters of the separate sections of the winding and it takes into account the existing non-linearities. The model thus created enables detection of random commutations in time of adjacent SCE switched over in series.

During the computer simulation, the following real conditions of commutations were considered:

- commutation at a random moment;
- commutation implemented by switching over in series of adjacent switches with SCE.

Account has been taken of the specifics of thyristor switches, whereby one commutation process comprises three intervals. The heaviest ADAVR mode has also been studies through simulation, in the first interval when switch K_3 is closed and all others are open; in the second interval when switches K_3 and K_4 are closed; and in the third interval when only switch K_4 is closed.

The system equations of the three intervals are represented in series for R-load (1), for L-load (2) and for C-load (3), whereby in the first interval the mesh current method is applied in a complex form: the system is solved by the Gauss-Jordan method; in the second and third interval the state variables method is applied and the systems are solved by the Runge–Kutta fourth-order method (RK4 method) in the temporal field.

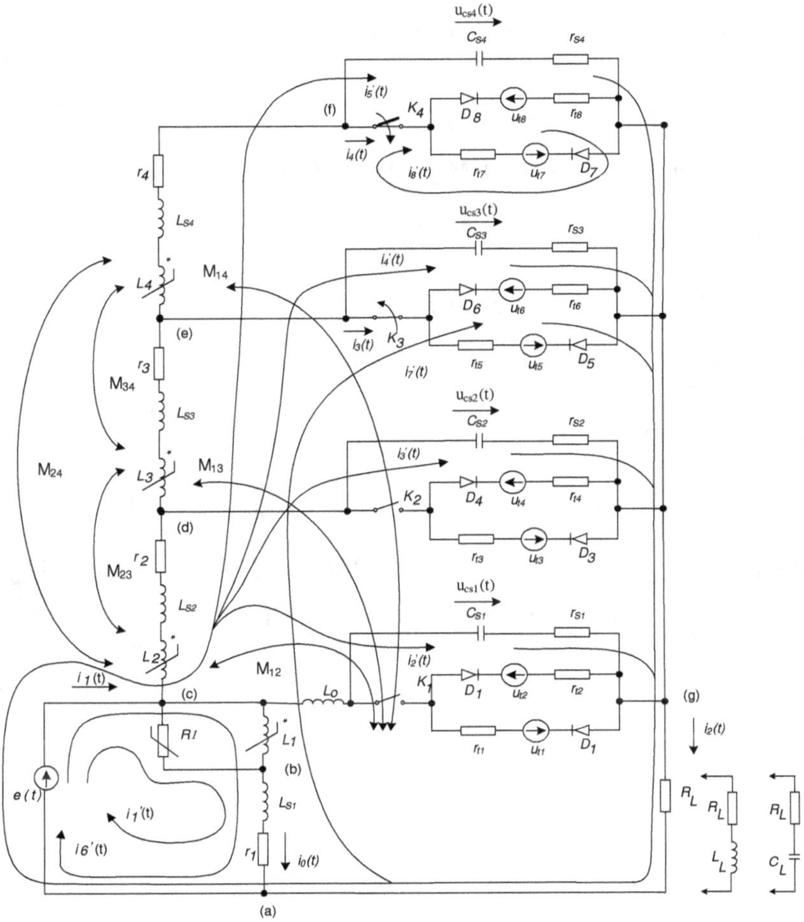

Fig.1. Equivalent circuit of ADAVR with four SCE.

First interval (established mode), mesh current method in a complex form for R-load

$$+(R_{\mu} + j\omega L_1)\dot{I}'_1 + j\omega L_1.\dot{I}'_2 - j\omega L_1.\dot{I}'_3 - j\omega(L_1 + M_{12})\dot{I}'_4 - j\omega\left(L_1 + M_{12} + M_{13}\right)\dot{I}'_5 -$$
$$-\omega\left(L_1 + M_{12} + M_{13} + M_{14}\right)\dot{I}'_6 - j\omega\left(L_1 + M_{12} + M_{13}\right)\dot{I}'_7 = 0$$

$$+j\omega L_1.\dot{I}'_1 + \left[r_1 + j\omega\left(L_{S1} + L_1\right)\right].\dot{I}'_2 - \left[r_1 + j\omega\left(L_{S1} + L_1\right)\right].\dot{I}'_3 - \left[r_1 + j\omega\begin{pmatrix} L_{S1} + L_1 + \\ +M_{12} \end{pmatrix}\right].\dot{I}'_4 -$$
$$-\left[r_1 + j\omega\left(L_{S1} + L_1 + M_{12} + M_{13}\right)\right].\dot{I}'_5 - \left[r_1 + j\omega\left(L_{S1} + L_1 + M_{12} + M_{13} + M_{14}\right)\right].\dot{I}'_6 -$$
$$-\left[r_1 + j\omega\left(L_{S1} + L_1 + M_{12} + M_{13}\right)\right].\dot{I}'_7 = \dot{E}$$

$$-j\omega L_1.\dot{I}'_1 - \left[r_1 + j\omega\left(L_{S1} + L_1\right)\right].\dot{I}'_2 + \left[r_1 + r_{S1} + R_T + j\omega\left(L_{S1} + L_1\right) - j\frac{1}{\omega C_{S1}}\right].\dot{I}'_3 +$$
$$+\left[r_1 + R_T + j\omega\left(L_{S1} + L_1 + M_{12}\right)\right].\dot{I}'_4 + \left[r_1 + R_T + j\omega\left(L_{S1} + L_1 + M_{12} + M_{13}\right)\right].\dot{I}'_5 +$$
$$+\left[r_1 + R_T + j\omega\left(L_{S1} + L_1 + M_{12} + M_{13} + M_{14}\right)\right].\dot{I}'_6 +$$
$$+\left[r_1 + R_T + j\omega\left(L_{S1} + L_1 + M_{12} + M_{13}\right)\right].\dot{I}'_7 = 0$$

$$-j\omega(L_1 + M_{12}).\dot{I}'_1 - \left[r_1 + j\omega\left(L_{S1} + L_1 + M_{12}\right)\right].\dot{I}'_2 +$$
$$+\left[r_1 + R_T + j\omega\left(L_{S1} + L_1 + M_{12}\right)\right].\dot{I}'_3 +$$
$$+\left[r_1 + r_2 + r_{S2} + R_T + j\omega\left(L_{S1} + L_1 + L_{S2} + L_2 + 2M_{12}\right) - j\frac{1}{\omega C_{S2}}\right].\dot{I}'_4 +$$
$$+\left[r_1 + r_2 + R_T + j\omega\left(L_{S1} + L_1 + L_{S2} + L_2 + 2M_{12} + M_{13} + M_{23}\right)\right].\dot{I}'_5 +$$
$$+\left[r_1 + r_2 + R_T + j\omega\left(L_{S1} + L_1 + L_{S2} + L_2 + 2M_{12} + M_{13} + M_{14} + M_{23} + M_{24}\right)\right].\dot{I}'_6 +$$
$$+\left[r_1 + r_2 + R_T + j\omega\left(L_{S1} + L_1 + L_{S2} + L_2 + 2M_{12} + M_{13} + M_{23}\right)\right].\dot{I}'_7 = 0$$

$$
\begin{aligned}
&-j\omega(L_1+M_{12}+M_{13}).\dot{I}_1^{'}-\left[r_1+j\omega\left(L_{S1}+L_1+M_{12}+M_{13}\right)\right].\dot{I}_2^{'}+\\
&+\left[r_1+R_T+j\omega\left(L_{S1}+L_1+M_{12}+M_{13}\right)\right].\dot{I}_3^{'}+\\
&+\left[r_1+r_2+R_T+j\omega\left(L_{S1}+L_1+L_{S2}+L_2+2M_{12}+M_{13}+M_{23}\right)\right].\dot{I}_4^{'}+\\
&+\left[\begin{array}{l} r_1+r_2+r_3+r_{S3}+R_T+j\omega\left(L_{S1}+L_1+L_{S2}+L_2+L_{S3}+L_3+2M_{12}+2M_{13}+2M_{23}\right)- \\ -j\dfrac{1}{\omega C_{S3}} \end{array}\right].\dot{I}_5^{'}+\\
&+\left[r_1+r_2+r_3+R_T+j\omega\left(\begin{array}{l} L_{S1}+L_1+L_{S2}+L_2+L_{S3}+L_3+2M_{12}+2M_{13}+M_{14}+ \\ +2M_{23}+M_{24}+M_{34} \end{array}\right)\right].\dot{I}_6^{'}+\\
&+\left[r_1+r_2+r_3+R_T+j\omega\left(L_{S1}+L_1+L_{S2}+L_2+L_{S3}+L_3+2M_{12}+2M_{13}+2M_{23}\right)\right].\dot{I}_7^{'}=0
\end{aligned}
$$

$$
\begin{aligned}
&-j\omega(L_1+M_{12}+M_{13}+M_{14}).\dot{I}_1^{'}-\left[r_1+j\omega\left(L_{S1}+L_1+M_{12}+M_{13}+M_{14}\right)\right].\dot{I}_2^{'}+\\
&+\left[r_1+R_T+j\omega\left(L_{S1}+L_1+M_{12}+M_{13}+M_{14}\right)\right].\dot{I}_3^{'}+\\
&+\left[r_1+r_2+R_T+j\omega\left(L_{S1}+L_1+L_{S2}+L_2+2M_{12}+M_{13}+M_{23}+M_{14}+M_{24}\right)\right].\dot{I}_4^{'}+\\
&+\left[r_1+r_2+r_3+R_T+j\omega\left(\begin{array}{l} L_{S1}+L_1+L_{S2}+L_2+L_{S3}+L_3+2M_{12}+2M_{13}+2M_{23}+ \\ +M_{14}+M_{24}+M_{34} \end{array}\right)\right].\dot{I}_5^{'}+\\
&+\left[\begin{array}{l} r_1+r_2+r_3+r_4+r_{S4}+R_T+\omega\left(\begin{array}{l} L_{S1}+L_1+L_{S2}+L_2+L_{S3}+L_3+L_{S4}+L_4+2M_{12}+ \\ +2M_{13}+2M_{14}+2M_{23}+2M_{24}+2M_{34} \end{array}\right)- \\ -j\dfrac{1}{\omega C_{S4}} \end{array}\right].\dot{I}_6^{'}+\\
&+\left[r_1+r_2+r_3+R_T+j\omega\left(\begin{array}{l} L_{S1}+L_1+L_{S2}+L_2+L_{S3}+L_3+2M_{12}+2M_{13}+2M_{23}+ \\ +M_{14}+M_{24}+M_{34} \end{array}\right)\right].\dot{I}_7^{'}=0
\end{aligned}
$$

$$-j\omega(L_1+M_{12}+M_{13}).\dot{I}_1^{'}-\left[r_1+j\omega\left(L_{S1}+L_1+M_{12}+M_{13}\right)\right].\dot{I}_2^{'}+$$
$$+\left[r_1+R_T+j\omega\left(L_{S1}+L_1+M_{12}+M_{13}\right)\right].\dot{I}_3^{'}+$$
$$+\left[r_1+r_2+R_T+j\omega\left(L_{S1}+L_1+L_{S2}+L_2+2M_{12}+M_{13}+M_{23}\right)\right].\dot{I}_4^{'}+$$
$$+\left[r_1+r_2+r_3+R_T+j\omega\left(L_{S1}+L_1+L_{S2}+L_2+L_{S3}+L_3+2M_{12}+2M_{13}+2M_{23}\right)\right].\dot{I}_5^{'}+$$
$$+\left[r_1+r_2+r_3+R_T+j\omega\begin{pmatrix}L_{S1}+L_1+L_{S2}+L_2+L_{S3}+L_3+2M_{12}+2M_{13}+M_{14}+\\+2M_{23}+M_{24}+M_{34}\end{pmatrix}\right].\dot{I}_6^{'}+$$
$$+\left[r_1+r_2+r_3+r_{t6}+R_T+j\omega\begin{pmatrix}L_{S1}+L_1+L_{S2}+L_2+L_{S3}+L_3+2M_{12}+2M_{13}+\\+2M_{23}\end{pmatrix}\right].\dot{I}_7^{'}=0$$

Second interval (first transient process), state variables method for R-load

$$+L_{S1}.\frac{di_1^{'}(t)}{dt}+0.\frac{di_2^{'}(t)}{dt}+0.\frac{di_3^{'}(t)}{dt}+0.\frac{di_4^{'}(t)}{dt}+0.\frac{di_5^{'}(t)}{dt}+L_{S1}.\frac{di_6^{'}(t)}{dt}-L_{S1}.\frac{di_7^{'}(t)}{dt}+0.\frac{di_8^{'}(t)}{dt}=$$
$$=-(r_1+R_\mu).i_1^{'}(t)+0.i_2^{'}(t)+0.i_3^{'}(t)+0.i_4^{'}(t)+0.i_5^{'}(t)-r_1.i_6^{'}(t)+r_1.i_7^{'}(t)+0.i_8^{'}(t)+e(t)$$

$$+0.\frac{di_1^{'}(t)}{dt}+L_0.\frac{di_2^{'}(t)}{dt}+0.\frac{di_3^{'}(t)}{dt}+0.\frac{di_4^{'}(t)}{dt}+0.\frac{di_5^{'}(t)}{dt}+0.\frac{di_6^{'}(t)}{dt}+0.\frac{di_7^{'}(t)}{dt}+0.\frac{di_8^{'}(t)}{dt}=$$
$$=0.i_1^{'}(t)-(r_{S1}+R_T).i_2^{'}(t)-R_T.i_3^{'}(t)-R_T.i_4^{'}(t)-R_T.i_5^{'}(t)+0.i_6^{'}(t)-R_T.i_7^{'}(t)+0.i_8^{'}(t)-u_{CS1}+e(t)$$

$$+0.\frac{di_1^{'}(t)}{dt}+0.\frac{di_2^{'}(t)}{dt}+(L_2+L_{S2}).\frac{di_3^{'}(t)}{dt}+(L_2+L_{S2}+M_{23}).\frac{di_4^{'}(t)}{dt}+(L_2+L_{S2}+M_{23}+M_{24}).\frac{di_5^{'}(t)}{dt}-$$
$$-M_{12}.\frac{di_6^{'}(t)}{dt}+(L_2+L_{S2}+M_{12}+M_{23}).\frac{di_8^{'}(t)}{dt}+M_{24}.\frac{di_8^{'}(t)}{dt}=0.i_1^{'}(t)-R_T.i_2^{'}(t)-(r_2+r_{S2}+R_T).i_3^{'}(t)-$$
$$-(r_2+R_T).i_4^{'}(t)-(r_2+R_T).i_5^{'}(t)+0.i_6^{'}(t)-(r_2+R_T).i_7^{'}(t)+0.i_8^{'}(t)-u_{CS2}(t)+e(t)$$

$$+0.\frac{di_1^{'}(t)}{dt}+0.\frac{di_2^{'}(t)}{dt}+(L_2+L_{S2}+M_{23}).\frac{di_3^{'}(t)}{dt}+(L_2+L_{S2}+L_3+L_{S3}+2.M_{23}).\frac{di_4^{'}(t)}{dt}+$$
$$+(L_2+L_{S2}+L_3+L_{s3}+2.M_{23}+M_{24}).\frac{di_5^{'}(t)}{dt}-(M_{12}+M_{13}).\frac{di_6^{'}(t)}{dt}+(L_2+L_{S2}+L_3+L_{S3}+$$
$$+2.M_{23}+M_{12}+M_{13}).\frac{di_8^{'}(t)}{dt}=0.i_1^{'}(t)-R_T.i_2^{'}(t)-(r_2+R_T).i_3^{'}(t)-(r_2+r_3+r_{S3}+R_T).i_4^{'}(t)-$$
$$-(r_2+r_3+R_T).i_5^{'}(t)+0.i_6^{'}(t)-(r_2+r_3+R_T).i_7^{'}(t)+0.i_8^{'}(t)-u_{CS3}(t)+e(t)$$

$$+0.\frac{di_1'(t)}{dt}+0.\frac{di_2'(t)}{dt}+(L_2+L_{S2}+M_{23}+M_{24}).\frac{di_3'(t)}{dt}+(L_2+L_{S2}+L_3+L_{S3}+2.M_{23}+$$

$$+M_{24}+M_{34}).\frac{di_4'(t)}{dt}+(L_2+L_{S2}+L_3+L_{S3}+L_4+L_{S4}+2.M_{23}+2.M_{34}+2.M_{24}).\frac{di_5'(t)}{dt}-$$

$$-(M_{12}+M_{13}+M_{14}).\frac{di_6'(t)}{dt}+(L_2+L_{S2}+L_3+L_{S3}+2.M_{23}+M_{24}+M_{34}+M_{12}+M_{13}+M_{14}).\frac{di_7'(t)}{dt}+$$

$$+(L_4+L_{S4}+M_{23}+M_{34}).\frac{di_8'(t)}{dt}=0.i_1'(t)-R_T.i_2'(t)-(r_2+R_T).i_3'(t)-(r_2+r_3+R_T).i_4'(t)-$$

$$-(r_2+r_3+r_4+r_{S4}+R_T).i_5'(t)+0.i_6'(t)-(r_2+r_3+R_T).i_7'(t)-r_4.i_8'(t)-u_{CS4}(t)+e(t)$$

$$+L_{S1}.\frac{di_1'(t)}{dt}+0.\frac{di_2'(t)}{dt}-M_{12}.\frac{di_3'(t)}{dt}-(M_{12}+M_{13}).\frac{di_4'(t)}{dt}-(M_{12}+M_{13}+M_{14}).\frac{di_5'(t)}{dt}+$$

$$+(L_1+L_{S1}).\frac{di_6'(t)}{dt}-(L_1+L_{S1}+M_{12}+M_{13}).\frac{di_7'(t)}{dt}-M_{14}.\frac{di_8'(t)}{dt}=-r_1.i_1'(t)+0.i_2'(t)+$$

$$+0.i_3'(t)+0.i_4'(t)+0.i_5'(t)-r_1.i_6'(t)+r_1.i_7'(t)+0.i_8'(t)+e(t)$$

$$-L_{S1}.\frac{di_1'(t)}{dt}+0.\frac{di_2'(t)}{dt}+(L_2+L_{S2}+M_{12}+M_{23}).\frac{di_3'(t)}{dt}+(L_2+L_{S2}+L_3+L_{S3}+2.M_{23}+$$

$$+M_{12}+M_{23}).\frac{di_4'(t)}{dt}+(L_2+L_{S2}+L_3+L_{S3}+2.M_{23}+M_{12}+M_{13}+M_{14}+M_{24}+M_{34}).\frac{di_5'(t)}{dt}-$$

$$-(L_1+L_{S1}+M_{12}+M_{13}).\frac{di_6'(t)}{dt}+(L_1+L_{S1}+L_2+L_{S2}+L_3+L_{S3}+2.M_{12}+2.M_{13}+2.M_{23}).\frac{di_7'(t)}{dt}+$$

$$+(M_{14}+M_{22}+M1_{23}).\frac{di_8'(t)}{dt}=+r_1.i_1'(t)-R_T.i_2'(t)-(r_2+R_T).i_3'(t)-(r_2+r_3+R_T).i_4'(t)-$$

$$-(r_2+r_3+R_T).i_5'(t)+r_1.i_6'(t)-(r_1+r_2+r_3+r_{t6(5)}+R_T).i_7'+r_{t6(5)}.i_8'(t)-sign\left[i_7'(t)\right].u_{t6}$$

$$+0.\frac{di_1'(t)}{dt}+0.\frac{di_2'(t)}{dt}+M_{24}.\frac{di_3'(t)}{dt}+(M_{22}+M_{34}).\frac{di_4'(t)}{dt}+(L_4+L_{S4}+M_{24}+M_{34}).\frac{di_5'(t)}{dt}-$$

$$-M_{14}.\frac{di_6'(t)}{dt}+(M_{14}+M_{24}+M_{34}).\frac{di_7'(t)}{dt}+(L_4+L_{S4}).\frac{di_8'(t)}{dt}=0.i_1'(t)+0.i_2'(t)+0.i_3'(t)+$$

$$+0.i_4'(t)-r_4.i_5'(t)+0.i_6'(t)+r_{t6(5)}.i_7'(t)-(r_4+r_{t6(5)}+r_{t8(7)}).i_8'(t)-sign\left[i_8'(t)\right].u_{t8}+sign.\left[i_7'(t)\right].u_{t6}$$

$$C_{S1}.\frac{du_{CS1}(t)}{dt}=i_2^{'}(t)$$

$$C_{S2}.\frac{du_{CS2}(t)}{dt}=i_3^{'}(t)$$

$$C_{S3}.\frac{du_{CS3}(t)}{dt}=i_4^{'}(t)$$

$$C_{S4}.\frac{du_{CS4}(t)}{dt}=i_5^{'}(t)$$

Third interval (second transient process), state variables method for R-load

$$+L_{S1}.\frac{di_1^{'}(t)}{dt}+0.\frac{di_2^{'}(t)}{dt}+0.\frac{di_3^{'}(t)}{dt}+0.\frac{di_4^{'}(t)}{dt}+0.\frac{di_5^{'}(t)}{dt}+L_{S1}.\frac{di_6^{'}(t)}{dt}-L_{S1}.\frac{di_8^{'}(t)}{dt}=$$
$$=-(r_1+r_\mu).i_1^{'}(t)+0.i_2^{'}(t)+0.i_3^{'}(t)+0.i_4^{'}(t)+0.i_5^{'}(t)-r_1.i_6^{'}(t)+r_1.i_8^{'}(t)+e(t)$$

$$+0.\frac{di_1^{'}(t)}{dt}+L_0.\frac{di_2^{'}(t)}{dt}+0.\frac{di_3^{'}(t)}{dt}+0.\frac{di_4^{'}(t)}{dt}+0.\frac{di_5^{'}(t)}{dt}+0.\frac{di_6^{'}(t)}{dt}+0.\frac{di_8^{'}(t)}{dt}=$$
$$=0.i_1^{'}(t)-(r_{S1}+R_T).i_2^{'}(t)-R_T.i_3^{'}(t)-R_T.i_4^{'}(t)-R_T.i_5^{'}(t)+0.i_6^{'}(t)-R_T.i_8^{'}(t)-u_{CS1}+e(t)$$

$$+0.\frac{di_1^{'}(t)}{dt}+0.\frac{di_2^{'}(t)}{dt}+(L_2+L_{S2}).\frac{di_3^{'}(t)}{dt}+(L_2+L_{S2}+M_{23}).\frac{di_4^{'}(t)}{dt}+(L_2+L_{S2}+M_{23}+M_{24}).\frac{di_5^{'}(t)}{dt}-$$
$$-M_{12}.\frac{di_6^{'}(t)}{dt}+(L_2+L_{S2}+M_{12}+M_{23}+M_{24}).\frac{di_8^{'}(t)}{dt}=0.i_1^{'}(t)-R_T.i_2^{'}(t)-(r_2+r_{S2}+R_T).i_3^{'}(t)-$$
$$-(r_2+R_T).i_4^{'}(t)-(r_2+R_T).i_5^{'}(t)+0.i_6^{'}(t)-(r_2+R_T).i_8^{'}(t)-u_{CS2}(t)+e(t)$$

$$+0.\frac{di_1^{'}(t)}{dt}+0.\frac{di_2^{'}(t)}{dt}+(L_2+L_{S2}+M_{23}).\frac{di_3^{'}(t)}{dt}+(L_2+L_{S2}+L_3+L_{S3}+2.M_{23}).\frac{di_4^{'}(t)}{dt}+$$
$$+(L_2+L_{S2}+L_3+L_{S3}+2.M_{23}+M_3+M_{24}).\frac{di_5^{'}(t)}{dt}-(M_{12}+M_{13}).\frac{di_6^{'}(t)}{dt}+$$
$$+(L_2+L_{S2}+L_3+L_{S3}+2.M_{23}+M_{12}+M_{13}+M_{24}+M_{34}).\frac{di_8^{'}(t)}{dt}=$$
$$=0.i_1^{'}(t)-R_T.i_2^{'}(t)--(r_2+R_T).i_3^{'}(t)-(r_2+r_3+r_{S3}+R_T).i_4^{'}(t)-(r_2+r_3+R_T).i_5^{'}(t)+$$
$$+0.i_6^{'}(t)-(r_2+r_3+R_T).i_8^{'}(t)-u_{CS3}(t)+e(t)$$

$$+0.\frac{di_1'(t)}{dt}+0.\frac{di_2'(t)}{dt}+(L_2+L_{S2}+M_{23}+M_{24}).\frac{di_3'(t)}{dt}+(L_2+L_{S2}+L_3+L_{S3}+2.M_{23}+$$
$$+M_{24}+M_{34}).\frac{di_4'(t)}{dt}+(L_2+L_{S2}+L_3+L_{S3}+L_4+L_{S4}+2.M_{23}+2.M_{24}+2.M_{34}).\frac{di_5'(t)}{dt}-$$
$$-(M_{12}+M_{13}+M_{14}).\frac{di_6'(t)}{dt}+(L_2+L_{S2}+L_3+L_{S3}+2.M_{23}+2.M_{24}+2.M_{34}+M_{12}+M_{13}+$$
$$+M_{14}+L_4+L_{S4}).\frac{di_8'(t)}{dt}=0.i_1'(t)-R_T.i_2'(t)-(r_2+R_T).i_3'(t)-(r_2+r_3+R_T).i_4'(t)-$$
$$-(r_2+r_3+r_4+r_{S4}+R_T).i_5'(t)+0.i_6'(t)-(r_2+r_3+r_4+R_T).i_8'(t)-u_{CS4}(t)+e(t)$$

$$+L_{S1}.\frac{di_1'(t)}{dt}+0.\frac{di_2'(t)}{dt}-M_{12}.\frac{di_3'(t)}{dt}-(M_{12}+M_{13}).\frac{di_4'(t)}{dt}-(M_{12}+M_{13}+M_{14}).\frac{di_5'(t)}{dt}+$$
$$+(L_1+L_{S1}).\frac{di_6'(t)}{dt}-(L_1+L_{S1}+M_{12}+M_{13}+M_{14}).\frac{di_8'(t)}{dt}=-r_1.i_1'(t)+0.i_2'(t)+0.i_3'(t)+$$
$$+0.i_4'(t)+0.i_5'(t)-r_1.i_6'(t)+r_1.i_8'(t)+e(t)$$

$$-L_{S1}.\frac{di_1'(t)}{dt}+0.\frac{di_2'(t)}{dt}+(L_2+L_{S2}+M_{12}+M_{23}+M_{24}).\frac{di_3'(t)}{dt}+(L_2+L_{S2}+L_3+L_{S3}+$$
$$+2.M_{23}+M_{12}+M_{24}+M_{34}).\frac{di_4'(t)}{dt}+(L_2+L_{S2}+L_3+L_{S3}+L_4+L_{S4}+2.M_{23}+M_{12}+$$
$$+M_{13}+M_{14}+2.M_{24}+2.M_{34}).\frac{di_5'(t)}{dt}-(L_1+L_{S1}+M_{12}+M_{13}+M_{14}).\frac{di_6'(t)}{dt}+$$
$$+(L_1+L_{S1}+L_2+L_{S2}+L_{S3}+2.M_{12}+2.M_{13}+2.M_{23}+L_4+L_{S4}+2.M_{14}+2.M_{24}+2.M_{34}).\frac{di_8'(t)}{dt}=$$
$$=+r_1.i_1'(t)-R_T.i_2'(t)-(r_2+R_T).i_3'(t)-(r_2+r_3+R_T).i_4'(t)-(r_2+r_3+r_4+R_T).i_5'(t)+r_1.i_6'(t)-$$
$$-(r_1+r_2+r_3+r_4+r_{t7(8)}+R_T).i_8'-sign\left[i_8'(t)\right].u_{t8(7)}$$

$$C_{S1}.\frac{du_{CS1}(t)}{dt}=i_2'(t)$$
$$C_{S2}.\frac{du_{CS2}(t)}{dt}=i_3'(t) \qquad (1)$$
$$C_{S3}.\frac{du_{CS3}(t)}{dt}=i_4'(t)$$
$$C_{S4}.\frac{du_{CS4}(t)}{dt}=i_5'(t)$$

First interval (established mode), mesh current method in a complex form for L-load

$$+(R_\mu+j\omega L_1)\dot{I}_1'+j\omega L_1.\dot{I}_2'-j\omega L_1.\dot{I}_3'-j\omega(L_1+M_{12})\dot{I}_4'-j\omega\left(L_1+M_{12}+M_{13}\right)\dot{I}_5'-$$
$$-j\omega\left(L_1+M_{12}+M_{13}+M_{14}\right)\dot{I}_6'-j\omega\left(L_1+M_{12}+M_{13}\right)\dot{I}_7'=0$$

$$+j\omega L_1.\dot{I}_1^{'}+\left[r_1+j\omega\left(L_{S1}+L_1\right)\right].\dot{I}_2^{'}-\left[r_1+j\omega\left(L_{S1}+L_1\right)\right].\dot{I}_3^{'}-$$
$$-\left[r_1+j\omega\left(L_{S1}+L_1+M_{12}\right)\right].\dot{I}_4^{'}-\left[r_1+j\omega\left(L_{S1}+L_1+M_{12}+M_{13}\right)\right].\dot{I}_5^{'}-$$
$$-\left[r_1+j\omega\left(L_{S1}+L_1+M_{12}+M_{13}+M_{14}\right)\right].\dot{I}_6^{'}-\left[r_1+j\omega\left(L_{S1}+L_1+M_{12}+M_{13}\right)\right].\dot{I}_7^{'}=\dot{E}$$

$$-j\omega L_1.\dot{I}_1^{'}-\left[r_1+j\omega\left(L_{S1}+L_1\right)\right].\dot{I}_2^{'}+\left[r_1+r_{S1}+R_T+j\omega\left(L_{S1}+L_1+L_T\right)-j\frac{1}{\omega C_{S1}}\right].\dot{I}_3^{'}+$$
$$+\left[r_1+R_T+j\omega\left(L_{S1}+L_1++L_T+M_{12}\right)\right].\dot{I}_4^{'}+\left[r_1+R_T+j\omega\left(L_{S1}+L_1+M_{12}+L_T+M_{13}\right)\right].\dot{I}_5^{'}+$$
$$+\left[r_1+R_T+j\omega\left(L_{S1}+L_1+L_T+M_{12}+M_{13}+M_{14}\right)\right].\dot{I}_6^{'}+\left[r_1+R_T+j\omega\begin{pmatrix}L_{S1}+L_1+L_T+\\+M_{12}+M_{13}\end{pmatrix}\right].\dot{I}_7^{'}=0$$

$$-j\omega(L_1+M_{12}).\dot{I}_1^{'}-\left[r_1+j\omega\left(L_{S1}+L_1+M_{12}\right)\right].\dot{I}_2^{'}+\left[r_1+R_T+j\omega\left(L_{S1}+L_1+L_T+M_{12}\right)\right].\dot{I}_3^{'}+$$
$$+\left[r_1+r_2+r_{S2}+R_T+j\omega\left(L_{S1}+L_1+L_{S2}+L_2+L_T+2M_{12}\right)-j\frac{1}{\omega C_{S2}}\right].\dot{I}_4^{'}+$$
$$+\left[r_1+r_2+R_T+j\omega\left(L_{S1}+L_1+L_{S2}+L_2+L_T+2M_{12}+M_{13}+M_{23}\right)\right].\dot{I}_5^{'}+$$
$$+\left[r_1+r_2+R_T+j\omega\left(L_{S1}+L_1+L_{S2}+L_2+L_T+2M_{12}+M_{13}+M_{14}+M_{23}+M_{24}\right)\right].\dot{I}_6^{'}+$$
$$+\left[r_1+r_2+R_T+j\omega\left(L_{S1}+L_1+L_{S2}+L_2+L_T+2M_{12}+M_{13}+M_{23}\right)\right].\dot{I}_7^{'}=0$$

$$-j\omega(L_1+M_{12}+M_{13}).\dot{I}_1^{'}-\left[r_1+j\omega\left(L_{S1}+L_1+M_{12}+M_{13}\right)\right].\dot{I}_2^{'}+$$
$$+\left[r_1+R_T+j\omega\left(L_{S1}+L_1+L_T+M_{12}+M_{13}\right)\right].\dot{I}_3^{'}+$$
$$+\left[r_1+r_2+R_T+j\omega\left(L_{S1}+L_1+L_{S2}+L_2+L_T+2M_{12}+M_{13}+M_{23}\right)\right].\dot{I}_4^{'}+$$
$$+\left[r_1+r_2+r_3+r_{S3}+R_T+j\omega\begin{pmatrix}L_{S1}+L_1+L_{S2}+L_2+L_{S3}+L_3+L_T+\\+2M_{12}+2M_{13}+2M_{23}\end{pmatrix}-j\frac{1}{\omega C_{S3}}\right].\dot{I}_5^{'}+$$
$$+\left[r_1+r_2+r_3+R_T+j\omega\begin{pmatrix}L_{S1}+L_1+L_{S2}+L_2+L_{S3}+L_3+L_T+2M_{12}+2M_{13}+\\+M_{14}+2M_{23}+M_{24}+M_{34}\end{pmatrix}\right].\dot{I}_6^{'}+$$
$$+\left[r_1+r_2+r_3+R_T+j\omega\left(L_{S1}+L_1+L_{S2}+L_2+L_{S3}+L_T+L_3+2M_{12}+2M_{13}+2M_{23}\right)\right].\dot{I}_7^{'}=0$$

$$-j\omega(L_1+M_{12}+M_{13}).\dot{I}_1^{'}-\left[r_1+j\omega\left(L_{S1}+L_1+M_{12}+M_{13}\right)\right].\dot{I}_2^{'}+$$

$$+\left[r_1+R_T+j\omega\left(L_{S1}+L_1+L_T+M_{12}+M_{13}\right)\right].\dot{I}_3^{'}+$$

$$+\left[r_1+r_2+R_T+j\omega\left(L_{S1}+L_1+L_{S2}+L_2+L_T+2M_{12}+M_{13}+M_{23}\right)\right].\dot{I}_4^{'}+$$

$$+\left[r_1+r_2+r_3+R_T+j\omega\left(L_{S1}+L_1+L_{S2}+L_2+L_{S3}+L_3+L_T+2M_{12}+2M_{13}+2M_{23}\right)\right].\dot{I}_5^{'}+$$

$$+\left[r_1+r_2+r_3+R_T+j\omega\begin{pmatrix}L_{S1}+L_1+L_{S2}+L_2+L_{S3}+L_3+L_T+2M_{12}+\\+2M_{13}+M_{14}+2M_{23}+M_{24}+M_{34}\end{pmatrix}\right].\dot{I}_6^{'}+$$

$$+\left[r_1+r_2+r_3+r_{i6}+R_T+j\omega\begin{pmatrix}L_{S1}+L_1+L_{S2}+L_2+L_{S3}+L_3+L_T+\\+2M_{12}+2M_{13}+2M_{23}\end{pmatrix}\right].\dot{I}_7^{'}=0$$

Second interval (first transient process), state variables method for L-load

$$+L_{S1}.\frac{di_1^{'}(t)}{dt}+0.\frac{di_2^{'}(t)}{dt}+0.\frac{di_3^{'}(t)}{dt}+0.\frac{di_4^{'}(t)}{dt}+0.\frac{di_5^{'}(t)}{dt}+L_{S1}.\frac{di_6^{'}(t)}{dt}-$$

$$-L_{S1}.\frac{di_7^{'}(t)}{dt}+0.\frac{di_8^{'}(t)}{dt}=-(r_1+R_\mu).i_1^{'}(t)+0.i_2^{'}(t)+0.i_3^{'}(t)+0.i_4^{'}(t)+$$

$$+0.i_5^{'}(t)-r_1.i_6^{'}(t)+r_1.i_7^{'}(t)+0.i_8^{'}(t)+e(t)$$

$$+0.\frac{di_1^{'}(t)}{dt}+(L_0+L_T).\frac{di_2^{'}(t)}{dt}+L_T.\frac{di_3^{'}(t)}{dt}+L_T.\frac{di_4^{'}(t)}{dt}+L_T.\frac{di_5^{'}(t)}{dt}+0.\frac{di_6^{'}(t)}{dt}+$$

$$+L_T.\frac{di_7^{'}(t)}{dt}+0.\frac{di_8^{'}(t)}{dt}=0.i_1^{'}(t)-(r_{S1}+R_T).i_2^{'}(t)-R_T.i_3^{'}(t)-R_T.i_4^{'}(t)-R_T.i_5^{'}(t)+$$

$$+0.i_6^{'}(t)-R_T.i_7^{'}(t)+0.i_8^{'}(t)-u_{CS1}+e(t)$$

$$+0.\frac{di_1^{'}(t)}{dt}+L_T.\frac{di_2^{'}(t)}{dt}+(L_2+L_{S2}+L_T).\frac{di_3^{'}(t)}{dt}+(L_2+L_{S2}+M_{23}+L_T).\frac{di_4^{'}(t)}{dt}+$$

$$+(L_2+L_{S2}+M_{23}+M_{24}+L_T).\frac{di_5^{'}(t)}{dt}-M_{12}.\frac{di_6^{'}(t)}{dt}++(L_2+L_{S2}+M_{12}+M_{23}+L_T).\frac{di_7^{'}(t)}{dt}+$$

$$+M_{24}.\frac{di_8^{'}(t)}{dt}=0.i_1^{'}(t)-R_T.i_2^{'}(t)-(r_2+r_{S2}+R_T).i_3^{'}(t)-(r_2+R_T).i_4^{'}(t)-(r_2+R_T).i_5^{'}(t)+$$

$$+0.i_6^{'}(t)-(r_2+R_T).i_7^{'}(t)+0.i_8^{'}(t)-u_{CS2}(t)+e(t)$$

$$+0.\frac{di_1'(t)}{dt}+L_T.\frac{di_2'(t)}{dt}+(L_2+L_{S2}+L_T+M_{23}).\frac{di_3'(t)}{dt}+(L_2+L_{S2}+L_3+$$
$$+L_{S3}+L_T+2.M_{23}).\frac{di_4'(t)}{dt}+(L_2+L_{S2}+L_3+L_{S3}+L_T+2.M_{23}+M_{34}+M_{24}).\frac{di_5'(t)}{dt}-$$
$$-(M_{12}+M_{13}).\frac{di_6'(t)}{dt}+(L_2+L_{S2}+L_3+L_{S3}+L_T+2.M_{23}+M_{12}+M_{13}).\frac{di_7'(t)}{dt}+$$
$$+(M_{24}+M_{33}).\frac{di_8'(t)}{dt}=0.i_1'(t)-R_T.i_2'(t)-(r_2+R_T).i_3'(t)-(r_2+r_3+r_{S3}+R_T).i_4'(t)-$$
$$-(r_2+r_3+R_T).i_5'(t)+0.i_6'(t)-(r_2+r_3+R_T).i_7'(t)+0.i_8'(t)-u_{CS3}(t)+e(t)$$

$$+0.\frac{di_1'(t)}{dt}+L_T.\frac{di_2'(t)}{dt}+(L_2+L_{S2}+L_T+M_{23}+M_{24}).\frac{di_3'(t)}{dt}+$$
$$+(L_2+L_{S2}+L_3+L_{S3}+L_T+2.M_{23}+M_{24}+M_{34}).\frac{di_4'(t)}{dt}+$$
$$(L_2+L_{S2}+L_3+L_{S3}+L_4+L_{S4}+L_T+2.M_{23}+2.M_{34}+2.M_{24}).\frac{di_5'(t)}{dt}-$$
$$-(M_{12}+M_{13}+M_{14}).\frac{di_6'(t)}{dt}+(L_2+L_{S2}+L_3+L_{S3}+L_T+2.M_{23}+M_{24}+$$
$$+M_{34}+M_{12}+M_{13}+M_{14}).\frac{di_7'(t)}{dt}+(L_4+L_{S4}+M_{23}+M_{34}).\frac{di_8'(t)}{dt}=0.i_1'(t)-$$
$$-R_T.i_2'(t)-(r_2+R_T).i_3'(t)-(r_2+r_3+R_T).i_4'(t)-(r_2+r_3+r_4+r_{S4}+R_T).i_5'(t)+$$
$$+0.i_6'(t)-(r_2+r_3+R_T).i_7'(t)-r_4.i_8'(t)-u_{CS4}(t)+e(t)$$

$$+L_{S1}.\frac{di_1'(t)}{dt}+0.\frac{di_2'(t)}{dt}-M_{12}.\frac{di_3'(t)}{dt}-(M_{12}+M_{13}).\frac{di_4'(t)}{dt}-(M_{12}+M_{13}+M_{14}).\frac{di_5'(t)}{dt}+$$
$$+(L_1+L_{S1}).\frac{di_6'(t)}{dt}-(L_1+L_{S1}+M_{12}+M_{13}).\frac{di_7'(t)}{dt}-M_{14}.\frac{di_8'(t)}{dt}=-r_1.i_1'(t)+0.i_2'(t)+$$
$$+0.i_3'(t)+0.i_4'(t)+0.i_5'(t)-r_1.i_6'(t)+r_1.i_7'(t)+0.i_8'(t)+e(t)$$

$$-L_{S1}.\frac{di_1^{'}(t)}{dt}+L_T.\frac{di_2^{'}(t)}{dt}+(L_2+L_{S2}+L_T+M_{12}+M_{23}).\frac{di_3^{'}(t)}{dt}+$$

$$+(L_2+L_{S2}+L_3+L_{S3}+L_T+2.M_{23}+M_{12}+M_{23}).\frac{di_4^{'}(t)}{dt}+$$

$$+(L_2+L_{S2}+L_3+L_{S3}+L_T+2.M_{23}+M_{12}+M_{13}+M_{14}+M_{24}+M_{34}).\frac{di_5^{'}(t)}{dt}-$$

$$-(L_1+L_{S1}+M_{12}+M_{13}).\frac{di_6^{'}(t)}{dt}++(L_1+L_{S1}+L_2+L_{S2}+L_3+L_{S3}+L_T+$$

$$+2.M_{12}+2.M_{13}+2.M_{23}).\frac{di_7^{'}(t)}{dt}+(M_{14}+M_{24}+M_{34}).\frac{di_8^{'}(t)}{dt}=+r_1.i_1^{'}(t)-R_T.i_2^{'}(t)-$$

$$-(r_2+R_T).i_3^{'}(t)-(r_2+r_3+R_T).i_4^{'}(t)-(r_2+r_3+R_T).i_5^{'}(t)+r_1.i_6^{'}(t)-(r_1+r_2+r_3+r_{t6(5)}+R_T).i_7^{'}+$$

$$+r_{t6(5)}.i_8^{'}(t)-sign\left[i_7^{'}(t)\right].u_{t6}\left[u_{CS3+rS3}(t)\right]$$

$$+0.\frac{di_1^{'}(t)}{dt}+0.\frac{di_2^{'}(t)}{dt}+M_{24}.\frac{di_3^{'}(t)}{dt}+(M_{22}+M_{34}).\frac{di_4^{'}(t)}{dt}+(L_4+L_{S4}+M_{24}+M_{34}).\frac{di_5^{'}(t)}{dt}-$$

$$-M_{14}.\frac{di_6^{'}(t)}{dt}+(M_{14}+M_{24}+M_{34}).\frac{di_7^{'}(t)}{dt}+(L_4+L_{S4}).\frac{di_8^{'}(t)}{dt}=0.i_1^{'}(t)+0.i_2^{'}(t)+0.i_3^{'}(t)+$$

$$+0.i_4^{'}(t)-r_4.i_5^{'}(t)+0.i_6^{'}(t)+r_{t6(5)}.i_7^{'}(t)-(r_4+r_{t6(5)}+r_{t8(7)}).i_8^{'}(t)-sign\left[i_8^{'}(t)\right].u_{t8}\left[u_{CS4+rS4}(t)\right]+$$

$$+sign\left[i_7^{'}(t)\right].u_{t6}\left[u_{CS3+rS4}(t)\right]$$

$$C_{S1}.\frac{du_{CS1}(t)}{dt}=i_2^{'}(t)$$

$$C_{S2}.\frac{du_{CS2}(t)}{dt}=i_3^{'}(t)$$

$$C_{S3}.\frac{du_{CS3}(t)}{dt}=i_4^{'}(t)$$

$$C_{S4}.\frac{du_{CS4}(t)}{dt}=i_5^{'}(t)$$

Third interval (second transient process), state variables method for L-load

$$+L_{S1}.\frac{di_1^{'}(t)}{dt}+0.\frac{di_2^{'}(t)}{dt}+0.\frac{di_3^{'}(t)}{dt}+0.\frac{di_4^{'}(t)}{dt}+0.\frac{di_5^{'}(t)}{dt}+L_{S1}.\frac{di_6^{'}(t)}{dt}-L_{S1}.\frac{di_8^{'}(t)}{dt}=$$

$$=-(r_1+r_\mu).i_1^{'}(t)+0.i_2^{'}(t)+0.i_3^{'}(t)+0.i_4^{'}(t)+0.i_5^{'}(t)-r_1.i_6^{'}(t)+r_1.i_8^{'}(t)+e(t)$$

$$+0.\frac{di_1^{'}(t)}{dt}+(L_0+L_T).\frac{di_2^{'}(t)}{dt}+L_T.\frac{di_3^{'}(t)}{dt}+L_T.\frac{di_4^{'}(t)}{dt}+L_T.\frac{di_5^{'}(t)}{dt}+0.\frac{di_6^{'}(t)}{dt}+L_T.\frac{di_8^{'}(t)}{dt}=$$

$$=0.i_1^{'}(t)-(r_{S1}+R_T).i_2^{'}(t)-R_T.i_3^{'}(t)-R_T.i_4^{'}(t)-R_T.i_5^{'}(t)+0.i_6^{'}(t)-R_T.i_8^{'}(t)-u_{CS1}+e(t)$$

$$+0.\frac{di_1'(t)}{dt}+L_T.\frac{di_2'(t)}{dt}+(L_2+L_{S2}+L_T).\frac{di_3'(t)}{dt}+(L_2+L_{S2}+L_T+M_{23}).\frac{di_4'(t)}{dt}+$$

$$+(L_2+L_{S2}+L_T+M_{23}+M_{24}).\frac{di_5'(t)}{dt}-M_{12}.\frac{di_6'(t)}{dt}+(L_2+L_{S2}+L_T+M_{12}+$$

$$+M_{23}+M_{24}).\frac{di_8'(t)}{dt}=0.i_1'(t)-R_T.i_2'(t)-(r_2+r_{S2}+R_T).i_3'(t)-(r_2+R_T).i_4'(t)-$$

$$-(r_2+R_T).i_5'(t)+0.i_6'(t)-(r_2+R_T).i_8'(t)-u_{CS2}(t)+e(t)$$

$$+0.\frac{di_1'(t)}{dt}+L_T.\frac{di_2'(t)}{dt}+(L_2+L_{S2}+L_T+M_{23}).\frac{di_3'(t)}{dt}+(L_2+L_{S2}+L_3+L_{S3}+$$

$$+L_T+2.M_{23}).\frac{di_4'(t)}{dt}+(L_2+L_{S2}+L_3+L_{s3}+L_T+2.M_{23}+M_{34}+M_{24}).\frac{di_5'(t)}{dt}-$$

$$-(M_{12}+M_{13}).\frac{di_6'(t)}{dt}+(L_2+L_{S2}+L_3+L_{S3}+2.M_{23}+M_{12}+M_{13}+L_T+M_{24}+M_{34}).\frac{di_8'(t)}{dt}=$$

$$=0.i_1'(t)-R_T.i_2'(t)-(r_2+R_T).i_3'(t)-(r_2+r_3+r_{S3}+R_T).i_4'(t)-(r_2+r_3+R_T).i_5'(t)+$$

$$+0.i_6'(t)-(r_2+r_3+R_T).i_8'(t)-u_{CS3}(t)+e(t)$$

$$+0.\frac{di_1'(t)}{dt}+L_T.\frac{di_2'(t)}{dt}+(L_2+L_{S2}+L_T+M_{23}+M_{24}).\frac{di_3'(t)}{dt}+(L_2+L_{S2}+L_3+L_{S3}+$$

$$+L_T+2.M_{23}+M_{24}+M_{34}).\frac{di_4'(t)}{dt}+L_2+L_{S2}+L_3+L_{S3}+L_4+L_{S4}+L_T+2.M_{23}+$$

$$+2.M_{24}+2.M_{34}).\frac{di_5'(t)}{dt}-(M_{12}+M_{13}+M_{14}).\frac{di_6'(t)}{dt}+(L_2+L_{S2}+L_3+$$

$$+L_{S3}+L_4+L_{S4}+L_T+2.M_{23}+2.M_{24}+2.M_{34}+M_{12}+M_{13}+M_{14}).\frac{di_8'(t)}{dt}=$$

$$=0.i_1'(t)-R_T.i_2'(t)-(r_2+R_T).i_3'(t)-(r_2+r_3+R_T).i_4'(t)-(r_2+r_3+r_4+r_{S4}+R_T).i_5'(t)+$$

$$+0.i_6'(t)-(r_2+r_3+r_4+R_T).i_8'(t)-u_{CS4}(t)+e(t)$$

$$+L_{S1}.\frac{di_1'(t)}{dt}+0.\frac{di_2'(t)}{dt}-M_{12}.\frac{di_3'(t)}{dt}-(M_{12}+M_{13}).\frac{di_4'(t)}{dt}-(M_{12}+M_{13}+M_{14}).\frac{di_5'(t)}{dt}+$$

$$+(L_1+L_{S1}).\frac{di_6'(t)}{dt}-(L_1+L_{S1}+M_{12}+M_{13}+M_{14}).\frac{di_8'(t)}{dt}=-r_1.i_1'(t)+0.i_2'(t)+0.i_3'(t)+$$

$$+0.i_4'(t)+0.i_5'(t)-r_1.i_6'(t)+r_1.i_8'(t)+e(t)$$

$$-L_{S1}.\frac{di_1'(t)}{dt}+L_T.\frac{di_2'(t)}{dt}+(L_2+L_{S2}+L_T+M_{12}+M_{23}+M_{24}).\frac{di_3'(t)}{dt}+$$

$$+(L_2+L_{S2}+L_3+L_{S3}+L_T+2.M_{23}+M_{12}+M_{24}+M_{34}).\frac{di_4'(t)}{dt}+$$

$$+(L_2+L_{S2}+L_3+L_{S3}+L_4+L_{S4}+L_T+2.M_{23}+M_{12}+M_{13}+M_{14}+2.M_{24}+2.M_{34}).\frac{di_5'(t)}{dt}-$$

$$-(L_1+L_{S1}+M_{12}+M_{13}+M_{14}).\frac{di_6'(t)}{dt}+(L_1+L_{S1}+L_2+L_{S2}+L_{S3}+L_T+2.M_{12}+2.M_{13}+$$

$$+2.M_{23}+L_4+L_{S4}+2.M_{14}+2.M_{24}+2.M_{34}).\frac{di_8'(t)}{dt}=+r_1.i_1'(t)-R_T.i_2'(t)-(r_2+R_T).i_3'(t)-$$

$$-(r_2+r_3+R_T).i_4'(t)-(r_2+r_3+r_4+R_T).i_5'(t)+r_1.i_6'(t)-(r_1+r_2+r_3+r_4+r_{t7(8)}+R_T).i_8'-sign\left[i_8'(t)\right].u_{t8(7)}$$

$$
\begin{aligned}
&C_{S1}.\frac{du_{CS1}(t)}{dt}=i_2'(t)\\
&C_{S2}.\frac{du_{CS2}(t)}{dt}=i_3'(t)\\
&C_{S3}.\frac{du_{CS3}(t)}{dt}=i_4'(t)\\
&C_{S4}.\frac{du_{CS4}(t)}{dt}=i_5'(t)
\end{aligned}
\qquad (2)
$$

First interval (established mode), mesh current method in a complex form for C-load

$$+(R_\mu+j\omega L_1)\dot{I}_1'+j\omega L_1.\dot{I}_2'-j\omega L_1.\dot{I}_3'-j\omega(L_1+M_{12})\dot{I}_4'-j\omega\left(L_1+M_{12}+M_{13}\right)\dot{I}_5'-$$
$$-j\omega\left(L_1+M_{12}+M_{13}+M_{14}\right)\dot{I}_6'-j\omega\left(L_1+M_{12}+M_{13}\right)\dot{I}_7'=0$$

$$+j\omega L_1.\dot{I}_1'+\left[r_1+j\omega\left(L_{S1}+L_1\right)\right].\dot{I}_2'-\left[r_1+j\omega\left(L_{S1}+L_1\right)\right].\dot{I}_3'-\left[r_1+j\omega\left(L_{S1}+L_1+M_{12}\right)\right].\dot{I}_4'-$$
$$-\left[r_1+j\omega\left(L_{S1}+L_1+M_{12}+M_{13}\right)\right].\dot{I}_5'-\left[r_1+j\omega\left(L_{S1}+L_1+M_{12}+M_{13}+M_{14}\right)\right].\dot{I}_6'-$$
$$-\left[r_1+j\omega\left(L_{S1}+L_1+M_{12}+M_{13}\right)\right].\dot{I}_7'=\dot{E}$$

$$-j\omega L_1.\dot{I}_1^{'}-\left[r_1+j\omega\left(L_{S1}+L_1\right)\right].\dot{I}_2^{'}+\begin{bmatrix} r_1+r_{S1}+R_T+j\omega\left(L_{S1}+L_1\right)- \\ -j(\frac{1}{\omega C_{S1}}+\frac{1}{\omega C_T}) \end{bmatrix}.\dot{I}_3^{'}+$$

$$+\left[r_1+R_T+j\omega\left(L_{S1}+L_1+M_{12}\right)-j\frac{1}{\omega C_T}\right].\dot{I}_4^{'}+\left[r_1+R_T+j\omega\begin{pmatrix} L_{S1}+L_1+ \\ +M_{12}+M_{13} \end{pmatrix}-j\frac{1}{\omega C_T}\right].\dot{I}_5^{'}+$$

$$+\left[r_1+R_T+j\omega\left(L_{S1}+L_1+M_{12}+M_{13}+M_{14}\right)-j\frac{1}{\omega C_T}\right].\dot{I}_6^{'}+$$

$$+\left[r_1+R_T+j\omega\left(L_{S1}+L_1+M_{12}+M_{13}-j\frac{1}{\omega C_T}\right)\right].\dot{I}_7^{'}=0$$

$$-j\omega(L_1+M_{12}).\dot{I}_1^{'}-\left[r_1+j\omega\left(L_{S1}+L_1+M_{12}\right)\right].\dot{I}_2^{'}+\left[r_1+R_T+j\omega\left(L_{S1}+L_1+M_{12}\right)-j\frac{1}{\omega C_T}\right].\dot{I}_3^{'}+$$

$$+\left[r_1+r_2+r_{S2}+R_T+j\omega\left(L_{S1}+L_1+L_{S2}+L_2+2M_{12}\right)-j(\frac{1}{\omega C_{S2}}+\frac{1}{\omega C_T})\right].\dot{I}_4^{'}+$$

$$+\left[r_1+r_2+R_T+j\omega\left(L_{S1}+L_1+L_{S2}+L_2+2M_{12}+M_{13}+M_{23}\right)-j\frac{1}{\omega C_T}\right].\dot{I}_5^{'}+$$

$$+\left[r_1+r_2+R_T+j\omega\left(L_{S1}+L_1+L_{S2}+L_2+2M_{12}+M_{13}+M_{14}+M_{23}+M_{24}\right)-j\frac{1}{\omega C_T}\right].\dot{I}_6^{'}+$$

$$+\left[r_1+r_2+R_T+j\omega\left(L_{S1}+L_1+L_{S2}+L_2+2M_{12}+M_{13}+M_{23}\right)-j\frac{1}{\omega C_T}\right].\dot{I}_7^{'}=0$$

$$-j\omega(L_1+M_{12}+M_{13}).\dot{I}_1^{'}-\left[r_1+j\omega\left(L_{S1}+L_1+M_{12}+M_{13}\right)\right].\dot{I}_2^{'}+$$

$$+\left[r_1+R_T+j\omega\left(L_{S1}+L_1+M_{12}+M_{13}-j\frac{1}{\omega C_T}\right)\right].\dot{I}_3^{'}+$$

$$+\left[r_1+r_2+R_T+j\omega\left(L_{S1}+L_1+L_{S2}+L_2+2M_{12}+M_{13}+M_{23}\right)-j\frac{1}{\omega C_T}\right].\dot{I}_4^{'}+$$

$$+\begin{bmatrix} r_1+r_2+r_3+r_{S3}+R_T+j\omega\left(L_{S1}+L_1+L_{S2}+L_2+L_{S3}+L_3+2M_{12}+2M_{13}+2M_{23}\right)- \\ -j(\frac{1}{\omega C_{S3}}+\frac{1}{\omega C_T}) \end{bmatrix}.\dot{I}_5^{'}+$$

$$+\left[r_1+r_2+r_3+R_T+j\omega\begin{pmatrix} L_{S1}+L_1+L_{S2}+L_2+L_{S3}+L_3+2M_{12}+2M_{13}+ \\ +M_{14}+2M_{23}+M_{24}+M_{34} \end{pmatrix}-j\frac{1}{\omega C_T}\right].\dot{I}_6^{'}+$$

$$+\left[r_1+r_2+r_3+R_T+j\omega\left(L_{S1}+L_1+L_{S2}+L_2+L_{S3}+L_3+2M_{12}+2M_{13}+2M_{23}\right)-j\frac{1}{\omega C_T}\right].\dot{I}_7^{'}=0$$

$$-j\omega(L_1+M_{12}+M_{13}+M_{14}).\dot{I}'_1-\left[r_1+j\omega\left(L_{S1}+L_1+M_{12}+M_{13}+M_{14}\right)\right].\dot{I}'_2+$$

$$+\left[r_1+R_T+j\omega\left(L_{S1}+L_1+M_{12}+M_{13}+M_{14}-j\frac{1}{\omega C_T}\right)\right].\dot{I}'_3+$$

$$+\left[r_1+r_2+R_T+j\omega\left(L_{S1}+L_1+L_{S2}+L_2+2M_{12}+M_{13}+M_{23}+M_{14}+M_{24}-j\frac{1}{\omega C_T}\right)\right].\dot{I}'_4+$$

$$+\left[r_1+r_2+r_3+R_T+j\omega\begin{pmatrix}L_{S1}+L_1+L_{S2}+L_2+L_{S3}+L_3+2M_{12}+2M_{13}+\\+2M_{23}+M_{14}+M_{24}+M_{34}-j\dfrac{1}{\omega C_T}\end{pmatrix}\right].\dot{I}'_5+$$

$$+\left[\begin{matrix}r_1+r_2+r_3+r_4+r_{S4}+R_T+j\omega\begin{pmatrix}L_{S1}+L_1+L_{S2}+L_2+L_{S3}+L_3+L_{S4}+L_4+2M_{12}+\\+2M_{13}+2M_{14}+2M_{23}+2M_{24}+2M_{34}\end{pmatrix}-\\-j\left(\dfrac{1}{\omega C_{S4}}+\dfrac{1}{\omega C_T}\right)\end{matrix}\right].\dot{I}'_6+$$

$$+\left[r_1+r_2+r_3+R_T+j\omega\begin{pmatrix}L_{S1}+L_1+L_{S2}+L_2+L_{S3}+L_3+2M_{12}+2M_{13}+\\+2M_{23}+M_{14}+M_{24}+M_{34}-j\dfrac{1}{\omega C_T}\end{pmatrix}\right].\dot{I}'_7=0$$

$$-j\omega(L_1+M_{12}+M_{13}).\dot{I}'_1-\left[r_1+j\omega\left(L_{S1}+L_1+M_{12}+M_{13}\right)\right].\dot{I}'_2+$$

$$+\left[r_1+R_T+j\omega\left(L_{S1}+L_1+M_{12}+M_{13}\right)-j\frac{1}{\omega C_T}\right].\dot{I}'_3+$$

$$+\left[r_1+r_2+R_T+j\omega\left(L_{S1}+L_1+L_{S2}+L_2+2M_{12}+M_{13}+M_{23}\right)-j\frac{1}{\omega C_T}\right].\dot{I}'_4+$$

$$+\left[r_1+r_2+r_3+R_T+j\omega\left(L_{S1}+L_1+L_{S2}+L_2+L_{S3}+L_3+2M_{12}+2M_{13}+2M_{23}\right)-j\frac{1}{\omega C_T}\right].\dot{I}'_5+$$

$$+\left[r_1+r_2+r_3+R_T+j\omega\begin{pmatrix}L_{S1}+L_1+L_{S2}+L_2+L_{S3}+L_3+2M_{12}+\\+2M_{13}+M_{14}+2M_{23}+M_{24}+M_{34}\end{pmatrix}\right].\dot{I}'_6+$$

$$+\left[r_1+r_2+r_3+r_{t6}+R_T+j\omega\begin{pmatrix}L_{S1}+L_1+L_{S2}+L_2+L_{S3}+L_3+\\+2M_{12}+2M_{13}+2M_{23}\end{pmatrix}-j\frac{1}{\omega C_T}\right].\dot{I}'_7=0$$

Second interval (first transient process), state variables method for C-load

$$+L_{S1}.\frac{di'_1(t)}{dt}+0.\frac{di'_2(t)}{dt}+0.\frac{di'_3(t)}{dt}+0.\frac{di'_4(t)}{dt}+0.\frac{di'_5(t)}{dt}+L_{S1}.\frac{di'_6(t)}{dt}-L_{S1}.\frac{di'_7(t)}{dt}+0.\frac{di'_8(t)}{dt}=$$
$$=-(r_1+R_\mu).i'_1(t)+0.i'_2(t)+0.i'_3(t)+0.i'_4(t)+0.i'_5(t)-r_1.i'_6(t)+r_1.i'_7(t)+0.i'_8(t)+e(t)$$

$$+0.\frac{di_1^{'}(t)}{dt}+L_0.\frac{di_2^{'}(t)}{dt}+0.\frac{di_3^{'}(t)}{dt}+0.\frac{di_4^{'}(t)}{dt}+0.\frac{di_5^{'}(t)}{dt}+0.\frac{di_6^{'}(t)}{dt}+0.\frac{di_7^{'}(t)}{dt}+0.\frac{di_8^{'}(t)}{dt}=$$
$$=0.i_1^{'}(t)-(r_{S1}+R_T).i_2^{'}(t)-R_T.i_3^{'}(t)-R_T.i_4^{'}(t)-R_T.i_5^{'}(t)+0.i_6^{'}(t)-R_T.i_7^{'}(t)+0.i_8^{'}(t)-u_{CS1}+$$
$$+u_{CT1}(t)+e(t)$$

$$+0.\frac{di_1^{'}(t)}{dt}+0.\frac{di_2^{'}(t)}{dt}+(L_2+L_{S2}).\frac{di_3^{'}(t)}{dt}+(L_2+L_{S2}+M_{23}).\frac{di_4^{'}(t)}{dt}+$$
$$+(L_2+L_{S2}+M_{23}+M_{24}).\frac{di_5^{'}(t)}{dt}-M_{12}.\frac{di_6^{'}(t)}{dt}+(L_2+L_{S2}+M_{12}+M_{23}).\frac{di_7^{'}(t)}{dt}+$$
$$+M_{24}.\frac{di_8^{'}(t)}{dt}=0.i_1^{'}(t)-R_T.i_2^{'}(t)-(r_2+r_{S2}+R_T).i_3^{'}(t)-(r_2+R_T).i_4^{'}(t)-(r_2+R_T).i_5^{'}(t)+$$
$$+0.i_6^{'}(t)-(r_2+R_T).i_7^{'}(t)+0.i_8^{'}(t)-u_{CS2}(t)+u_{CT1}(t)+e(t)$$

$$+0.\frac{di_1^{'}(t)}{dt}+0.\frac{di_2^{'}(t)}{dt}+(L_2+L_{S2}+M_{23}).\frac{di_3^{'}(t)}{dt}+(L_2+L_{S2}+L_3+L_{S3}+2.M_{23}).\frac{di_4^{'}(t)}{dt}+$$
$$+(L_2+L_{S2}+L_3+L_{S3}+2.M_{23}+M_{34}+M_{24}).\frac{di_5^{'}(t)}{dt}-(M_{12}+M_{13}).\frac{di_6^{'}(t)}{dt}+$$
$$+(L_2+L_{S2}+L_3+L_{S3}+2.M_{23}+M_{12}+M_{13}).\frac{di_8^{'}(t)}{dt}=0.i_1^{'}(t)-R_T.i_2^{'}(t)-(r_2+R_T).i_3^{'}(t)-$$
$$-(r_2+r_3+r_{S3}+R_T).i_4^{'}(t)-(r_2+r_3+R_T).i_5^{'}(t)+0.i_6^{'}(t)-(r_2+r_3+R_T).i_7^{'}(t)+0.i_8^{'}(t)-$$
$$-u_{CS3}(t)+e(t)$$

$$+0.\frac{di_1^{'}(t)}{dt}+0.\frac{di_2^{'}(t)}{dt}+(L_2+L_{S2}+M_{23}+M_{24}).\frac{di_3^{'}(t)}{dt}+(L_2+L_{S2}+L_3+L_{S3}+$$
$$+2.M_{23}+M_{24}+M_{34}).\frac{di_4^{'}(t)}{dt}+(L_2+L_{S2}+L_3+L_{S3}+L_4+L_{S4}+2.M_{23}+2.M_{34}+$$
$$+2.M_{24}).\frac{di_5^{'}(t)}{dt}-(M_{12}+M_{13}+M_{14}).\frac{di_6^{'}(t)}{dt}+(L_2+L_{S2}+L_3+L_{S3}+2.M_{23}+M_{24}+$$
$$+M_{34}+M_{12}+M_{13}+M_{14}).\frac{di_7^{'}(t)}{dt}+(L_4+L_{S4}+M_{23}+M_{34}).\frac{di_8^{'}(t)}{dt}=0.i_1^{'}(t)-R_T.i_2^{'}(t)-$$
$$-(r_2+R_T).i_3^{'}(t)-(r_2+r_3+R_T).i_4^{'}(t)-(r_2+r_3+r_4+r_{S4}+R_T).i_5^{'}(t)+0.i_6^{'}(t)-$$
$$-(r_2+r_3+R_T).i_7^{'}(t)-r_4.i_8^{'}(t)-u_{CS4}(t)+u_{CT}(t)+e(t)$$

$$-L_{S1}.\frac{di_1'(t)}{dt}+0.\frac{di_2'(t)}{dt}+(L_2+L_{S2}+M_{12}+M_{23}).\frac{di_3'(t)}{dt}+(L_2+L_{S2}+L_3+L_{S3}+$$

$$+2.M_{23}+M_{12}+M_{23}).\frac{di_4'(t)}{dt}+(L_2+L_{S2}+L_3+L_{S3}+2.M_{23}+M_{12}+M_{13}+M_{14}+$$

$$+M_{24}+M_{34}).\frac{di_5'(t)}{dt}-(L_1+L_{S1}+M_{12}+M_{13}).\frac{di_6'(t)}{dt}+(L_1+L_{S1}+L_2+L_{S2}+L_3+L_{S3}+$$

$$+2.M_{12}+2.M_{13}+2.M_{23}).\frac{di_7'(t)}{dt}+(M_{14}+M_{24}+M_{34}).\frac{di_8'(t)}{dt}=+r_1.i_1'(t)-$$

$$-R_T.i_2'(t)-(r_2+R_T).i_3'(t)-(r_2+r_3+R_T).i_4'(t)-(r_2+r_3+R_T).i_5'(t)+r_1.i_6'(t)-$$

$$-(r_1+r_2+r_3+r_{t6(5)}+R_T).i_7'+r_{t6(5)}.i_8'(t)+u_{CT}(t)-sign\left[i_7'(t)\right].u_{t6}\left[u_{CS3+rS3}(t)\right]$$

$$+0.\frac{di_1'(t)}{dt}+0.\frac{di_2'(t)}{dt}+M_{24}.\frac{di_3'(t)}{dt}+(M_{22}+M_{34}).\frac{di_4'(t)}{dt}+(L_4+L_{S4}+M_{24}+M_{34}).\frac{di_5'(t)}{dt}-$$

$$-M_{14}.\frac{di_6'(t)}{dt}+(M_{14}+M_{24}+M_{34}).\frac{di_7'(t)}{dt}+(L_4+L_{S4}).\frac{di_8'(t)}{dt}=0.i_1'(t)+0.i_2'(t)+0.i_3'(t)+$$

$$+0.i_4'(t)-r_4.i_5'(t)+0.i_6'(t)+r_{t6(5)}.i_7'(t)-(r_4+r_{t6(5)}+r_{t8(7)}).i_8'(t)-sign\left[i_8'(t)\right].u_{t8}\left[u_{CS4+rS4}(t)\right]+$$

$$+sign\left[i_7'(t)\right].u_{t6}\left[u_{CS3+rS4}(t)\right]$$

$$C_{S1}.\frac{du_{CS1}(t)}{dt}=i_2'(t)$$

$$C_{S2}.\frac{du_{CS2}(t)}{dt}=i_3'(t)$$

$$C_{S3}.\frac{du_{CS3}(t)}{dt}=i_4'(t)$$

$$C_{S4}.\frac{du_{CS4}(t)}{dt}=i_5'(t)$$

$$C_T.\frac{du_{CT}(t)}{dt}=i_2'(t)+i_3'(t)+i_4'(t)+i_5'(t)+i_7'(t)$$

Third interval (second transient process), state variables method for C-load

$$+L_{S1}.\frac{di_1'(t)}{dt}+0.\frac{di_2'(t)}{dt}+0.\frac{di_3'(t)}{dt}+0.\frac{di_4'(t)}{dt}+0.\frac{di_5'(t)}{dt}+L_{S1}.\frac{di_6'(t)}{dt}-L_{S1}.\frac{di_8'(t)}{dt}=$$

$$=-(r_1+r_\mu).i_1'(t)+0.i_2'(t)+0.i_3'(t)+0.i_4'(t)+0.i_5'(t)-r_1.i_6'(t)+r_1.i_8'(t)+e(t)$$

$$+0.\frac{di_1'(t)}{dt}+L_0.\frac{di_2'(t)}{dt}+0.\frac{di_3'(t)}{dt}+0.\frac{di_4'(t)}{dt}+0.\frac{di_5'(t)}{dt}+0.\frac{di_6'(t)}{dt}+0.\frac{di_8'(t)}{dt}=0.i_1'(t)-$$

$$-(r_{S1}+R_T).i_2'(t)-R_T.i_3'(t)-R_T.i_4'(t)-R_T.i_5'(t)+0.i_6'(t)-R_T.i_8'(t)-u_{CS1}+u_{CT}(t)+e(t)$$

$$+0.\frac{di_1^{'}(t)}{dt}+0.\frac{di_2^{'}(t)}{dt}+(L_2+L_{S2}).\frac{di_3^{'}(t)}{dt}+(L_2+L_{S2}+M_{23}).\frac{di_4^{'}(t)}{dt}+$$
$$+(L_2+L_{S2}+M_{23}+M_{24}).\frac{di_5^{'}(t)}{dt}-M_{12}.\frac{di_6^{'}(t)}{dt}++(L_2+L_{S2}+M_{12}+M_{23}+M_{24}).\frac{di_8^{'}(t)}{dt}=$$
$$=0.i_1^{'}(t)-R_T.i_2^{'}(t)-(r_2+r_{S2}+R_T).i_3^{'}(t)-(r_2+R_T).i_4^{'}(t)-(r_2+R_T).i_5^{'}(t)+0.i_6^{'}(t)-$$
$$-(r_2+R_T).i_8^{'}(t)-u_{CS2}(t)+u_{CT}(t)+e(t)$$

$$+0.\frac{di_1^{'}(t)}{dt}+0.\frac{di_2^{'}(t)}{dt}+(L_2+L_{S2}+M_{23}).\frac{di_3^{'}(t)}{dt}+(L_2+L_{S2}+L_3+L_{S3}+2.M_{23}).\frac{di_4^{'}(t)}{dt}+$$
$$+(L_2+L_{S2}+L_3+L_{S3}+2.M_{23}+M_{34}+M_{24}).\frac{di_5^{'}(t)}{dt}-(M_{12}+M_{13}).\frac{di_6^{'}(t)}{dt}+$$
$$+(L_2+L_{S2}+L_3+L_{S3}+2.M_{23}+M_{12}+M_{13}+M_{24}+M_{34}).\frac{di_8^{'}(t)}{dt}=0.i_1^{'}(t)-R_T.i_2^{'}(t)-$$
$$-(r_2+R_T).i_3^{'}(t)-(r_2+r_3+r_{S3}+R_T).i_4^{'}(t)-(r_2+r_3+R_T).i_5^{'}(t)+0.i_6^{'}(t)-(r_2+r_3+R_T).i_8^{'}(t)-$$
$$-u_{CS3}(t)+u_{CT}(t)+e(t)$$

$$+0.\frac{di_1^{'}(t)}{dt}+0.\frac{di_2^{'}(t)}{dt}+(L_2+L_{S2}+M_{23}+M_{24}).\frac{di_3^{'}(t)}{dt}+(L_2+L_{S2}+L_3+L_{S3}+$$
$$+2.M_{23}+M_{24}+M_{34}).\frac{di_4^{'}(t)}{dt}+(L_2+L_{S2}+L_3+L_{S3}+L_4+L_{S4}+2.M_{23}+$$
$$+2.M_{24}+2.M_{34}).\frac{di_5^{'}(t)}{dt}-(M_{12}+M_{13}+M_{14}).\frac{di_6^{'}(t)}{dt}+(L_2+L_{S2}+L_3+$$
$$+L_{S3}+2.M_{23}+2.M_{24}+2.M_{34}+M_{12}+M_{13}+M_{14}+L_4+L_{S4}).\frac{di_8^{'}(t)}{dt}=$$
$$=0.i_1^{'}(t)-R_T.i_2^{'}(t)-(r_2+R_T).i_3^{'}(t)-(r_2+r_3+R_T).i_4^{'}(t)-(r_2+r_3+r_4+r_{S4}+R_T).i_5^{'}(t)+$$
$$+0.i_6^{'}(t)-(r_2+r_3+r_4+R_T).i_8^{'}(t)-u_{CS4}(t)+u_{CT}(t)+e(t)$$

$$+L_{S1}.\frac{di_1^{'}(t)}{dt}+0.\frac{di_2^{'}(t)}{dt}-M_{12}.\frac{di_3^{'}(t)}{dt}-(M_{12}+M_{13}).\frac{di_4^{'}(t)}{dt}-(M_{12}+M_{13}+M_{14}).\frac{di_5^{'}(t)}{dt}+$$
$$+(L_1+L_{S1}).\frac{di_6^{'}(t)}{dt}-(L_1+L_{S1}+M_{12}+M_{13}+M_{14}).\frac{di_8^{'}(t)}{dt}=-r_1.i_1^{'}(t)+0.i_2^{'}(t)+0.i_3^{'}(t)+$$
$$+0.i_4^{'}(t)+0.i_5^{'}(t)-r_1.i_6^{'}(t)+r_1.i_8^{'}(t)+e(t)$$

$$-L_{S1}.\frac{di_1^{'}(t)}{dt}+0.\frac{di_2^{'}(t)}{dt}+(L_2+L_{S2}+M_{12}+M_{23}+M_{24}).\frac{di_3^{'}(t)}{dt}+(L_2+L_{S2}+L_3+L_{S3}+$$
$$+2.M_{23}+M_{12}+M_{24}+M_{34}).\frac{di_4^{'}(t)}{dt}+(L_2+L_{S2}+L_3+L_{S3}+L_4+L_{S4}+2.M_{23}+M_{12}+$$
$$+M_{13}+M_{14}+2.M_{24}+2.M_{34}).\frac{di_5^{'}(t)}{dt}-(L_1+L_{S1}+M_{12}+M_{13}+M_{14}).\frac{di_6^{'}(t)}{dt}+$$
$$+(L_1+L_{S1}+L_2+L_{S2}+L_{S3}+2.M_{12}+2.M_{13}+2.M_{23}+L_4+L_{S4}+2.M_{14}+2.M_{24}+$$
$$+2.M_{34}).\frac{di_8^{'}(t)}{dt}=+r_1.i_1^{'}(t)-R_T.i_2^{'}(t)-(r_2+R_T).i_3^{'}(t)-(r_2+r_3+R_T).i_4^{'}(t)-$$
$$-(r_2+r_3+r_4+R_T).i_5^{'}(t)+r_1.i_6^{'}(t)-(r_1+r_2+r_3+r_4+r_{t7(8)}+R_T).i_8^{'}+u_{CT}(t)-sign\left[i_8^{'}(t)\right].u_{t8(7)}$$

$$C_{S1}.\frac{du_{CS1}(t)}{dt}=i_2^{'}(t)$$
$$C_{S2}.\frac{du_{CS2}(t)}{dt}=i_3^{'}(t)$$
$$C_{S3}.\frac{du_{CS3}(t)}{dt}=i_4^{'}(t) \qquad (3)$$
$$C_{S4}.\frac{du_{CS4}(t)}{dt}=i_5^{'}(t)$$
$$C_T.\frac{du_{CT}(t)}{dt}=i_2^{'}(t)+i_3^{'}(t)+i_4^{'}(t)+i_5^{'}(t)+i_8^{'}(t)$$

Fig. 2, 3 and 4 show the graphs of the circuit for the three intervals of the commutation process.

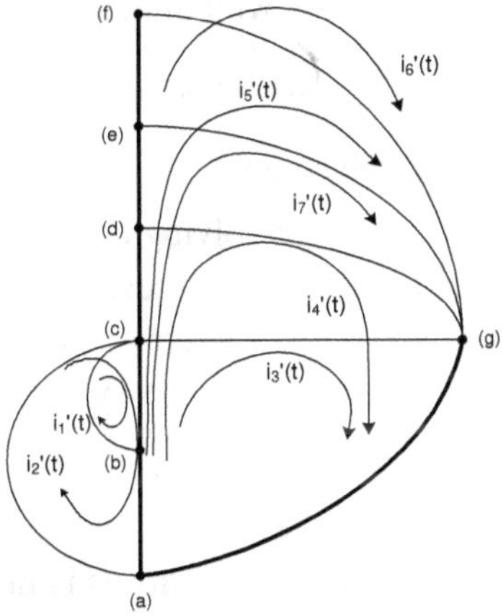

Fig. 2. Graph of the circuit for the first interval.

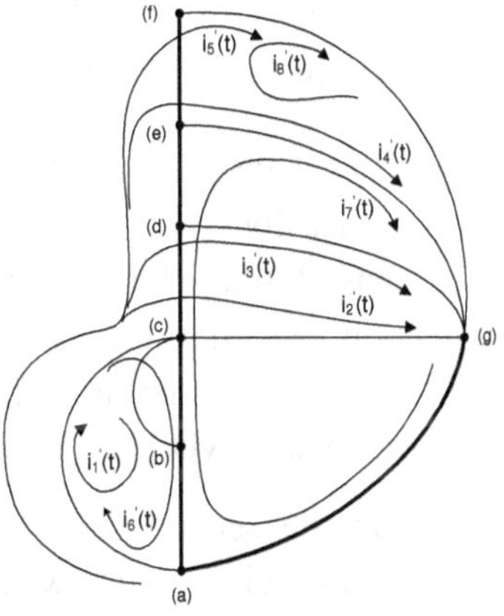

Fig. 3. Graph of the circuit for the second interval.

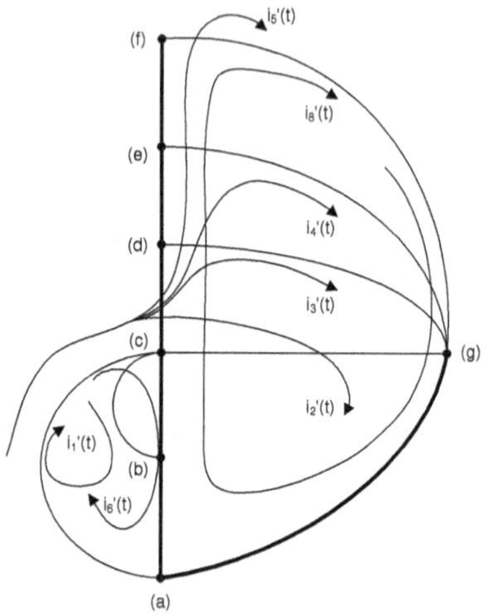

Fig. 4. Graph of the circuit for the third interval.

In the first interval, ADAVR still operates in an established alternating-current mode, where the electrical balance is described by the mesh current method in a complex form. The number of the independent loops is seven and the system of equations consists of seven equations with seven unknowns. The general form of the system assumes the following matrix form:

$$[Y_1].[\dot{I}'] = [\dot{E}], \tag{4}$$

where $[Y_1]$ is a complex conductivity matrix of ADAVR with dimensions 7 rows and 7 columns (i.e. 7x7); $[\dot{I}']$ is the matrix column of mesh currents $\dot{I}_1'; \dot{I}_2';...\dot{I}_7'$ with dimensions 7x1; $[\dot{E}]$ is the matrix-column of the complexes of the mesh emf with dimensions 7x1. The numerical method, chosen for solving equation (4) is the Gauss-Jordan method, and the matrix $[Y_1]$ is not singular, regardless of the selection version of the independent loops of the circuit. Therefore, there are no particular difficulties to determine the initial response of ADAVR until the moment when switch K_4 gets closed.

In the second interval, the first transitive process in ADAVR can be observed providing that switches K_3 and K_4 are closed. The analysis is made by the state variables method, where eight of the state variables are mesh currents $i_1'(t); i_2'(t);...;i_7'(t);i_8'(t)$, and the other four are the voltages over the switch-off capacitors $u_{cs1}(t);\ldots; u_{cs4}(t)$. The analysis is made in the temporal field, whereby the system of equations contains twelve equations with twelve unknowns. The matrix of the system of equations is almost singular. This leads to error accumulation during the computations for some specific values of the commutation phase of switch K_4. No alteration of the independent loops in the various versions results in improvement of the matrix type of the system of equations, i.e. it remains almost singular. Nonetheless, the independent loops are chosen in the form shown in Fig. 1 and Fig. 3. In this situation, the general matrix form of the system of equations is as follows:

$$[A_2]\frac{d}{dt}[x(t)] = [B_2].[x(t)] + [e_2(t)], \tag{5}$$

where the matrices $[A_2]$ and $[B_2]$ have dimensions 12x12 and consist of nine submatrices:

$$[A_2] = \begin{bmatrix} A_{11} & A_{12} & A_{13} \\ \hline A_{21} & A_{22} & A_{23} \\ \hline A_{31} & A_{32} & A_{33} \end{bmatrix}, \quad [B_2] = \begin{bmatrix} B_{11} & B_{12} & B_{13} \\ \hline B_{21} & B_{22} & B_{23} \\ \hline B_{31} & B_{32} & B_{33} \end{bmatrix}. \tag{6}$$

The matrices $[x(t)]$ and $[e_2(t)]$ have dimensions 12x1 and each of them consists of three submatrices:

$$[x(t)] = \begin{bmatrix} X_{11}(t) \\ \overline{X_{21}(t)} \\ \overline{X_{31}(t)} \end{bmatrix}, \quad [e_2(t)] = \begin{bmatrix} E_{11}(t) \\ \overline{E_{21}(t)} \\ \overline{E_{31}(t)} \end{bmatrix}. \tag{7}$$

Here, matrix $[x(t)]$ contains the state variables in the order as listed above, and matrix $[e_2(t)]$ contains the input emf e(t) and the voltage drops u_{t5}, u_{t6}, u_{t7} and u_{t8} over the thyristor switches K_3 and K_4 in a saturated state.

The dimensions of the submatrices are as follows:
$[A_{11}]$ and $[B_{11}]$ are 5x5; $[A_{12}]$ and $[B_{12}]$ are 5x3; $[A_{13}]$ and $[B_{13}]$ are 5x4;
$[A_{21}]$ and $[B_{21}]$ are 3x5; $[A_{22}]$ and $[B_{22}]$ are 3x3; $[A_{23}]$ and $[B_{23}]$ are 3x4;
$[A_{31}]$ and $[B_{31}]$ are 4x5; $[A_{32}]$ and $[B_{32}]$ are 4x3; $[A_{33}]$ and $[B_{33}]$ are 4x4;
$[X_{11}(t)]$ and $[E_{11}(t)]$ are 5x1; $[X_{21}(t)]$ and $[E_{21}(t)]$ are 3x1; $[X_{31}(t)]$ and $[E_{31}(t)]$ are 4x1.

The matrices $[A_2]$ and $[B_2]$ are sparse, i.e. they have over 30% zero elements, and the following submatrices are also zero ones: $[A_{13}]$; $[A_{23}]$; $[A_{31}]$; $[A_{32}]$; $[B_{23}]$; $[B_{31}(:,1)]$; $[B_{32}]$; $[B_{33}]$; $[E_{31}(t)]$. In addition, submatrices $[A_{33}]$, $[B_{13}]$, $[B_{31}]$ are diagonal. The submatrices of the state variables have the following form:

$$[X_{11}(t)] = \begin{bmatrix} i_1'(t) \\ i_2'(t) \\ \dots\dots \\ i_5'(t) \end{bmatrix}; \; [X_{21}(t)] = \begin{bmatrix} i_6'(t) \\ i_7'(t) \\ i_8'(t) \end{bmatrix}; \; [X_{31}(t)] = \begin{bmatrix} u_{cs1}(t) \\ u_{cs2}(t) \\ u_{cs3}(t) \\ u_{cs4}(t) \end{bmatrix}. \qquad (8)$$

The algorithm, used in order to avoid the fact that matrix $[A_2]$ is almost singular, consists in breaking down equation (5) into three matrix equations containing the submatrices listed above, and solving them one after another. With each step forward in time, the following matrix equations are solved:

$$1)\; [A_{22}]\frac{d}{dt}[X_{21}(t)] = [B_{22}].[X_{21}(t)] + [E_{21}(t)], \qquad (9)$$

whereby equation (9) is multiplied on the left and right by the opposite matrix $[A_{22}]^{-1}$ and then the approximate values of the first derivatives $\frac{d}{dt}[X_{21}(t)]$ are substituted in the next matrix equation:

$$2)[A_{11}]\frac{d}{dt}[X_{11}(t)] = \left[B_{11} \middle| B_{12} \middle| B_{13}\right].[X(t)] - [A_{12}]\frac{d}{dt}[X_{21}(t)] + [E_{11}(t)], \qquad (10)$$

whereby equation (10) is multiplied on the left and right with the opposite matrix $[A_{11}]^{-1}$ and then RK4 method is applied. The obtained solution is substituted in the next matrix equation:

$$3)\; [A_{33}]\frac{d}{dt}[X_{31}(t)] = [B_{31}(:;2:5)].[X_{11}(t)], \qquad (11)$$

whereby equation (11) is multiplied on the left and right by the opposite matrix $[A_{33}]^{-1}$ and then RK4 method is applied. The obtained solution is substituted in the last matrix equation:

$$4)[A_{22}]\frac{d}{dt}[X_{21}(t)] = -[A_{21}]\frac{d}{dt}[X_{11}(t)] + \left[B_{21} \middle| B_{22}\right].\begin{bmatrix} X_{11}(t) \\ \hline X_{21}(t) \end{bmatrix} + [E_{11}(t)], \qquad (12)$$

whereby equation (12) is multiplied on the left and right by the opposite matrix $[A_{22}]^{-1}$ and then RK4 method is applied. The obtained solution is substituted in equation (9) so that the next step forward in time be performed, and then again in equations (10), (11) and (12), until the second interval is finished and switch K_3 opens.

The computation procedure operates with non-singular submatrices $[A_{22}]$, $[A_{11}]$ and $[A_{33}]$ and the obtained solutions do not accumulate error in the time of the second subinterval.

In the third interval, the second transitive process can be observed; here, however, only switch K_4 is closed. The analysis is made in a similar way by the state variables method in the time domain. The number of the state variables is now eleven since here the mesh current $i_7^{'}(t)$ is removed from the group of unknowns in the previous interval. The system of equations in third interval has the following form:

$$[A_3]\frac{d}{dt}[x(t)] = [B_3].[x(t)] + [e_3(t)], \tag{13}$$

where the matrices $[A_3]$ and $[B_3]$ have dimensions 11x11 and each of them consists of three submatrices:

$$[A_3] = \left[\begin{array}{c|c|c} a_{11} & a_{12} & a_{13} \\ \hline a_{21} & a_{22} & a_{23} \\ \hline a_{31} & a_{32} & a_{33} \end{array}\right], \quad [B_3] = \left[\begin{array}{c|c|c} b_{11} & b_{12} & b_{13} \\ \hline b_{21} & b_{22} & b_{23} \\ \hline b_{31} & b_{32} & b_{33} \end{array}\right]. \tag{14}$$

The matrices $[x(t)]$ and $[e_3(t)]$ have dimensions 11x1 and each of them consists of three submatrices:

$$[x(t)] = \begin{bmatrix} x_{11}(t) \\ \overline{x_{21}(t)} \\ \overline{x_{31}(t)} \end{bmatrix}, \quad [e_3(t)] = \begin{bmatrix} e_{11}(t) \\ \overline{e_{21}(t)} \\ \overline{e_{31}(t)} \end{bmatrix}. \tag{15}$$

The matrix $[x(t)]$ contains the state variables in the order given for the second interval without the current $i_7^{'}(t)$, and the matrix $[e_3(t)]$ contains the input emf $e(t)$ and the drops u_{t7} and u_{t8} over the thyristor switch K_4 in a saturated state.

The dimensions of the submatrices are as follows:
$[a_{11}]$ and $[b_{11}]$ are 5x5; $[a_{12}]$ and $[b_{12}]$ are 5x2; $[a_{13}]$ and $[b_{13}]$ are 5x4;
$[a_{21}]$ and $[b_{21}]$ are 2x5; $[a_{22}]$ and $[b_{22}]$ are 2x2; $[a_{23}]$ and $[b_{23}]$ are 2x4;
$[a_{31}]$ and $[b_{31}]$ are 4x5; $[a_{32}]$ and $[b_{32}]$ are 4x2; $[a_{33}]$ and $[b_{33}]$ are 4x4;
$[x_{11}(t)]$ and $[e_{11}(t)]$ are 5x1; $[x_{21}(t)]$ and $[e_{21}(t)]$ are 2x1; $[x_{31}(t)]$ and $[e_{31}(t)]$ are 4x1.

The matrices $[A_3]$ and $[B_3]$ are sparse, while matrix $[A_3]$ is almost singular. The following submatrices are also zero ones: $[a_{13}]$; $[a_{23}]$; $[a_{31}]$; $[a_{32}]$; $[b_{23}]$; $[b_{32}]$;

$[b_{33}]$; $[e_{31}(t)]$. In addition, submatrices $[a_{33}]$, $[b_{13}]$, $[b_{31}]$ are diagonal. The submatrices of the state variables have the following form:

$$[x_{11}(t)] = \begin{bmatrix} i_1^{'}(t) \\ i_2^{'}(t) \\ \ldots\ldots \\ i_5^{'}(t) \end{bmatrix}; \ [x_{21}(t)] = \begin{bmatrix} i_6^{'}(t) \\ i_8^{'}(t) \end{bmatrix}; \ [x_{31}(t)] = \begin{bmatrix} u_{cs1}(t) \\ u_{cs2}(t) \\ u_{cs3}(t) \\ u_{cs4}(t) \end{bmatrix}. \quad (16)$$

The algorithm, used in order to avoid the fact that matrix $[A_3]$ is almost singular, consists in breaking down equation (13) into three matrix equations containing the submatrices listed above, and at each step forward in time, the following matrix equations are solved:

$$1)\ [a_{22}]\frac{d}{dt}[x_{21}(t)] = [b_{22}].[x_{21}(t)] + [e_{21}(t)], \quad (17)$$

whereby equation (17) is multiplied on the left and right by the opposite matrix $[a_{22}]^{-1}$ and then the obtained approximate values for the first derivatives $\frac{d}{dt}[x_{21}(t)]$ are substituted in the next matrix equation:

$$2)\ [a_{11}]\frac{d}{dt}[x_{11}(t)] = \left[b_{11} | b_{12} | b_{13}\right].[x(t)] - [a_{12}]\frac{d}{dt}[x_{21}(t)] + [e_{11}(t)], \quad (18)$$

whereby equation (18) is multiplied on the left and right by the opposite matrix $[a_{11}]^{-1}$ and then RK4 method is applied. The obtained solution is substituted in the next matrix equation:

$$3)\ [a_{33}]\frac{d}{dt}[x_{31}(t)] = [b_{31}].[x_{11}(t)], \quad (19)$$

whereby equation (19) is multiplied on the left and right with the opposite matrix $[a_{33}]^{-1}$ and then RK4 method is applied to the normalized equation. The obtained solution is substituted in the last matrix equation:

$$4)\ [a_{22}]\frac{d}{dt}[x_{21}(t)] = -[a_{21}]\frac{d}{dt}[x_{11}(t)] + \left[b_{21} | b_{22}\right].\begin{bmatrix} x_{11}(t) \\ \overline{x_{21}(t)} \end{bmatrix} + [e_{21}(t)], \quad (20)$$

whereby equation (20) is multiplied on the left and right with the opposite matrix $[a_{22}]^{-1}$ and then RK4 method is applied.

The procedure can be repeated for the next step forward by using equations (17), (18), (19) and (20) again in the same sequence.

In the third interval, the computation procedure operates with non-singular matrices $[a_{22}]$, $[a_{11}]$ and $[a_{33}]$, and the obtained solutions do not accumulate computation error in time.

The algorithm in the second and third interval is identical. It starts with approximate computation of a group of the first derivatives of the state variables (equations (9) and (17)) and then the accurate solutions are obtained by soling three subgroups of equations one after another (equations (10), (8), (11) and (18), (19), (20) respectively), which were obtained through breaking down parts of the general system and of ADAVR (equations (5) and (13)), which should not be solved directly since the matrices of systems ($[A_2]$ and $[A_3]$) are almost singular. The stability of the solutions, obtained via this algorithm, does not depend on the commutation phase of switches K_3 and K_4.

The automated computer program AVTO was developed for simulation of processes in ADAVR with SCE in the environment of the computer program MATLAB in view of its suitability to perform complex mathematical computations in a matrix form and its features allowing spatial visualization in graphic form of the obtained numerical results from the studied parameters. Fig. 5 shows a flow-chart of the program AVTO.

The computer program AVTO contains an input block for the parameters of the computation process and the parameters of the studied ADAVR. This block is for setting the duration of the first and the third interval of the commutation process, the size of the computation step, the effective value of input voltage, as well as the parameters of the windings, ferromagnetic core, commuting elements, switch-off assemblies and the load. The phase of the desired commutation is set and the inherent parameters of the commutation process are computed.

The block for determining the processes in the first interval of the commutation process is used to make analysis by the mesh current method in a complex form, whereby the Gauss-Jordan method is applied. After that the effective values of the studied parameters are determined and they are visualized in the temporal field.

The block for determining the initial conditions of the second interval of the commutation process is used to set a correct assignment of the problem for the first transitive process in ADAVR.

Next is the block for determining the processes in the second interval of the commutation process or the so-called first transitive process. Here, the analysis is made by help of the state variables method in the temporal field by applying a specialized algorithm to avoid the fact that the matrix of the system control of ADAVR is almost singular. In the computation procedures, RK4 method is used. At the same time, the general control system is broken down into three subsystems which are solved one after another [7].

Next is the block for visualization of the studied parameters in the second interval and the block for determining the initial conditions in the third interval of the commutation process.

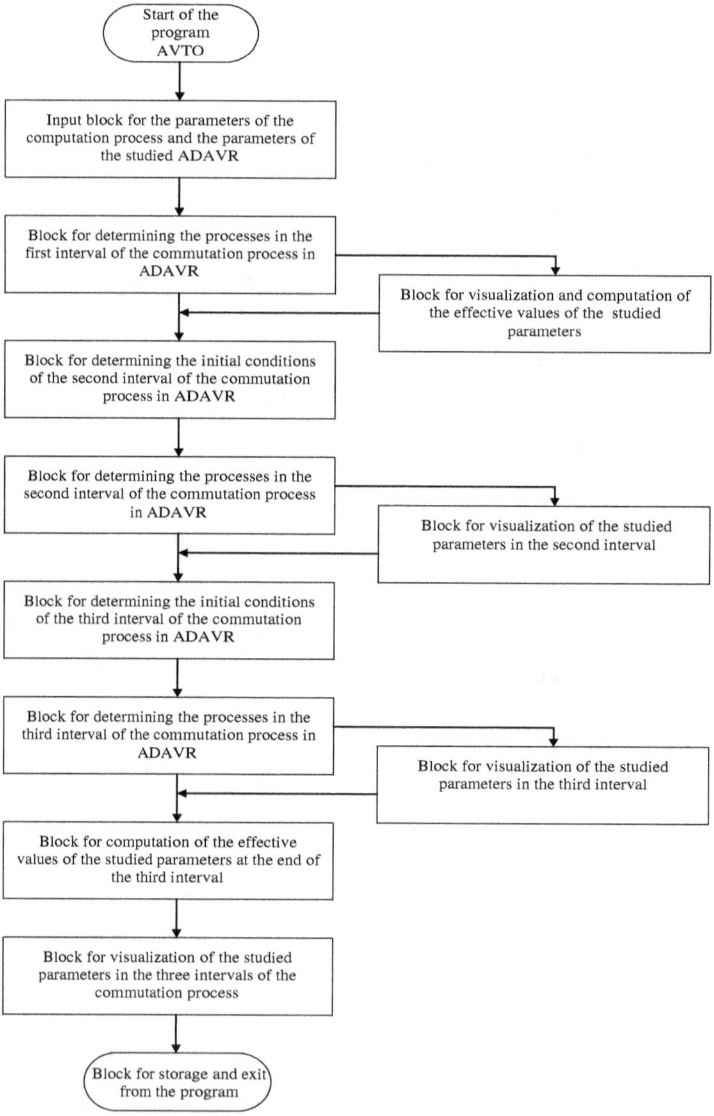

Fig. 5. Flow-chart of the program AVTO.

The block for determining the processes in third interval is used to study the so-called second transitive process. Here the analysis is made by the state variables method in the temporal field with the help of an algorithm similar to the one used for analysis of the first transitive process. The reason is again the fact that the matrix of the system is almost singular. The system of equations is broken down into subsystems which are solved by RK4 method. After that the solutions are visualized in third interval, followed by computation of the effective values of the studied parameters after the second transitive process has died away and visualization of all studied parameters in the three intervals of the commutation process.

Finally, the obtained numerical and graphic results are stored and the program is exited.

Fig. 6 shows the charts of the studied parameters in the three intervals of the commutation process of ADAVR with 0,78kW of resistive load. The commutation phase of switch K_4 is selected to be φ=225°. Fig. 6a shows the input current $i_1(t)$ as function of time; Fig. 6b shows the output current $i_2(t)$; Fig. 6c – current $i_0(t)$ in the main section of ADAVR; Fig. 6d – current $i_3(t)$ through switch K_3; Fig. 6e – current $i_4(t)$ through switch K_4; Fig. 6f – voltage $u_{cs3+rs3}(t)$ over the third switch-off group $C_{S3}+r_{S3}$; Fig. 6g – voltage $u_{cs4+rs4}(t)$ over the fourth switch-off group $C_{S4}+r_{S4}$; Fig. 6h – current $i_{cs3+rs3}(t)$ through the third switch-off group; Fig. 6i – current $i_{cs4+rs4}(t)$ through the fourth switch-off group.

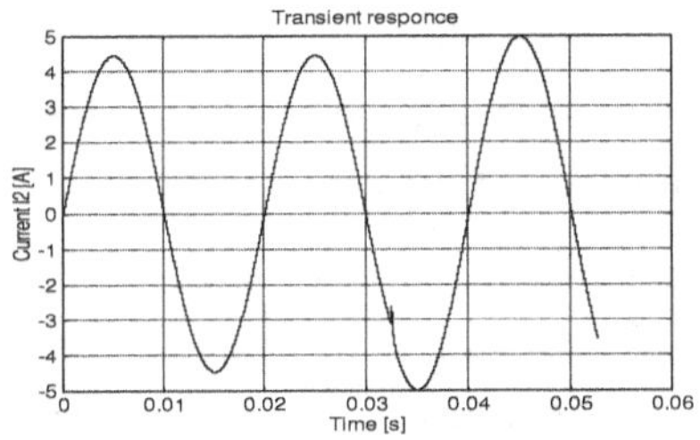

a)

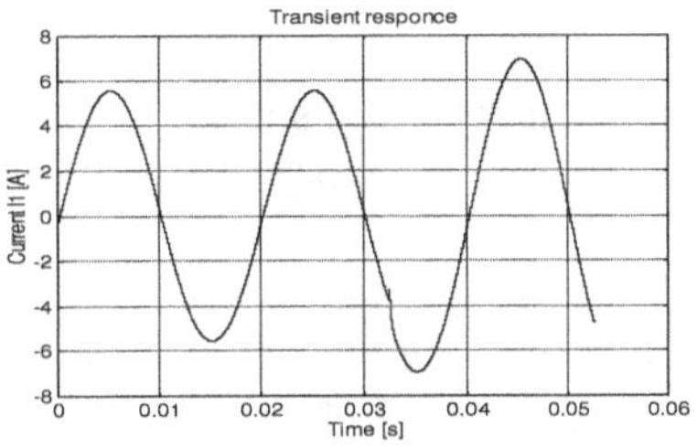

b)

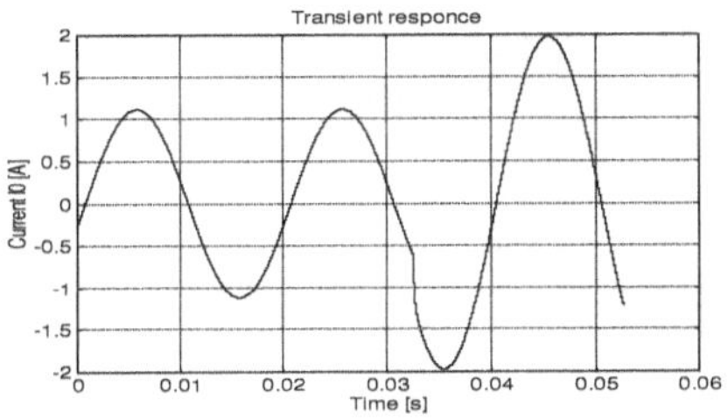

c)

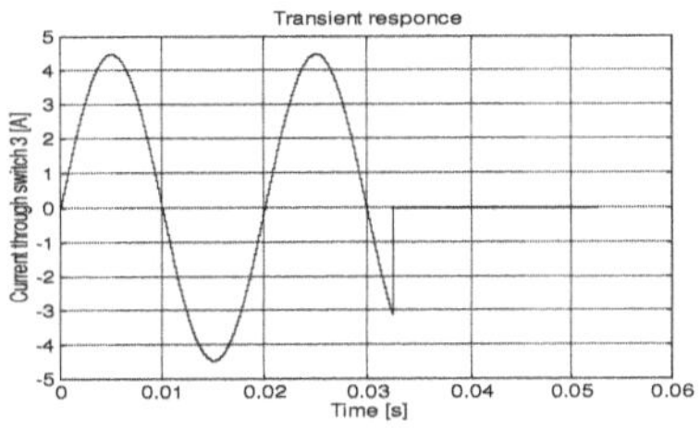

d)

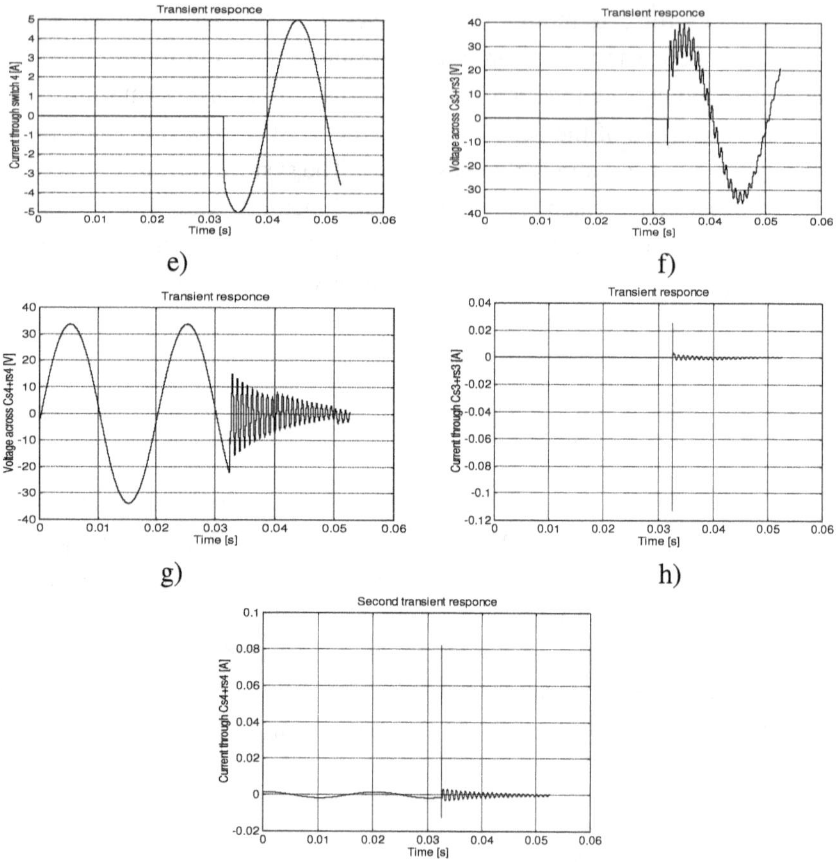

i)

Fig. 6. Charts of the studied parameters in ADAVR with resistive load for the three intervals of the commutation process.

Fig. 7 shows the relationship between the first transitive process duration and the commutation phase of switch K_4. The results were obtained for three resistive loads each of respective power as follows: 0,42kW – marked by asterisk; 0,78kW – marked by circles; 1,35kW – marked by crosses.

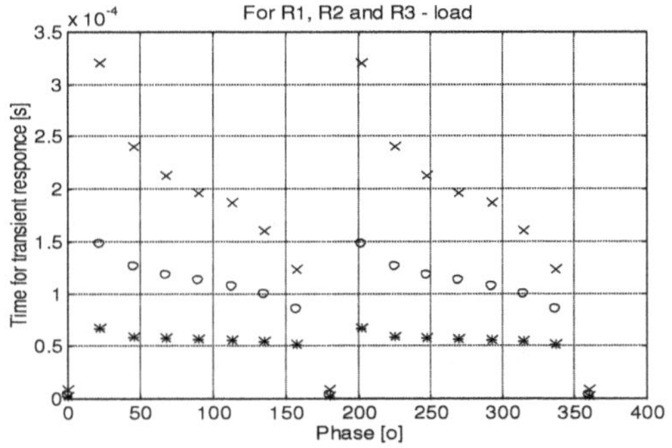

Fig. 7. Diagram of the relationship between the first transitive process duration, the commutation phase of switch K_4 and the power of the load.

Fig. 8 shows charts of the relationships between the amplitude values of the studied parameters in ADAVR during the first transitive process, the commutation phase of switch K_4 and the same three resistive loads with identical markings. Fig. 8a shows input current i_{1max}; Fig. 8b shows output current i_{2max}; Fig. 8c – current i_{0max} in the main section of ADAVR; Fig. 8d – current i_{3max} through switch K_3; Fig. 8e – current i_{4max} through switch K_4.

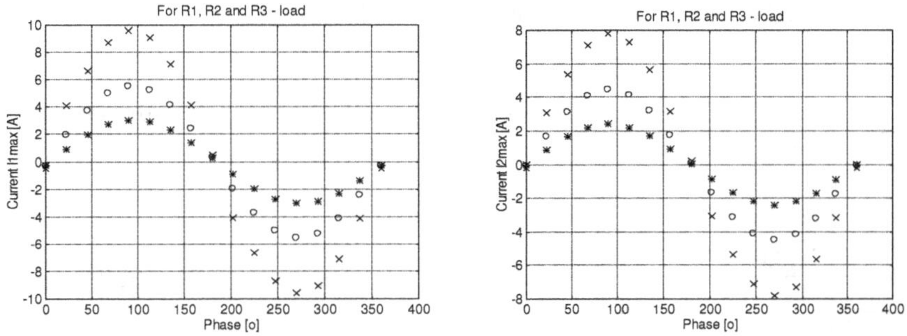

a) b)

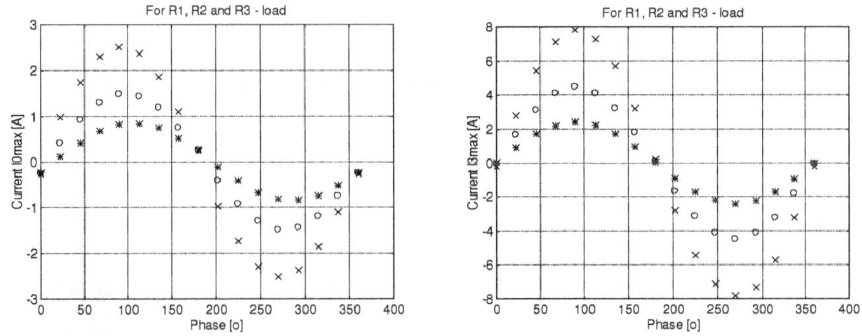

c) d)

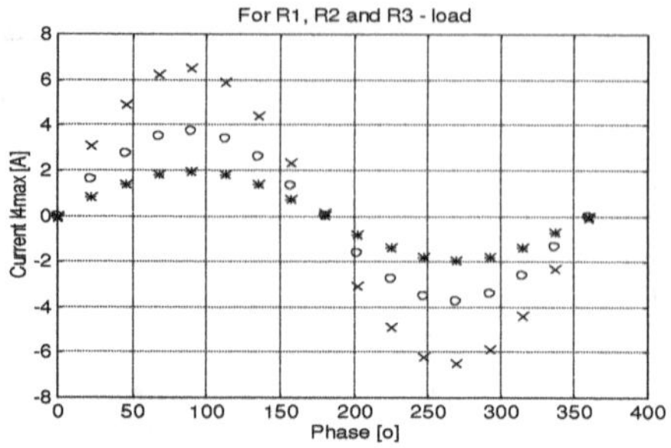

e)

Fig.8. Diagrams of the relationships between the amplitude values of the parameters in ADAVR during the first transitive process, the commutation phase of switch K_4 and the power of the load.

The obtained graphic results very well correspond to the experimental data from a number of tests conducted for ADAVR with four SCE and the same three resistive loads [7].

Table 1 shows data from the experimental testing and computer simulations by help of the proposed algorithm and the program AVTO for a particular ADAVR at input voltage of 160V and three different resistive loads.

Table 1. Results from a number of tests and computer simulations.

	Results from physical experiment					***Results from computer simulations***				
Load	U_1	I_1	I_0	$I_2 \equiv I_T$	$U_2 \equiv U_T$	U_1	I_1	I_0	$I_2 \equiv I_T$	$U_2 \equiv U_T$
№	V	A	A	A	V	V	A	A	A	A
1	160	2,69	0,81	1,9	220	160	2,6674	0,7934	1,8865	218,436
2	160	4,94	1,42	3,56	220	160	4,9232	1,3990	3,5313	218,230
3	160	8,45	2,38	6,16	219	160	8,4757	2,3551	6,1247	217,740

The data indicate very good correspondence with high precision of the computed response of ADAVR related to the results from the physical measurements.

3. CONCLUSIONS

The following conclusions can be drawn from the set forth hereinabove:

1. A mathematical model is proposed for description of the processes of commutation of the terminals of an autotransformer in a discrete alternating voltage regulator as well as a detailed algorithm for solving the system of

equations describing the electrical balance in ADAVR, which allows to overcome the singularity of the matrix of the system in the second and third interval during the action of the voltage regulator.

2. A computer program AVTO has been developed for simulation of the complex processes occurring in ADAVR during the commutation processes, which enables a thorough analysis of the said devices.

3. The developed computer program comprises the specifics of the circuit design, the alteration of the parameters of the autotransformer as a function of the mode, the size and character of the load, the initial phase of commutation and the parameters of the used semiconductor switches. This enables a more comprehensive quantitative analysis of the character and duration of the commutation process, as well as of the electric loading of the individual circuits and semiconductor switches.

4. The data from a number of computer simulations and physical tests have been compared for a large range of loads, indicating very good correspondence between the obtained results.

5. A system version is proposed for virtual design of discrete voltage regulators, whose application would accelerate these processes and economize financial means in the process of development of such devices.

6. The practical implementation of discrete alternating voltage regulators connected before one load or a group of loads is multivalent and could refer to parameter adjustment of the main as well as to improvement of the power and functional indicators of loads connected after a regulator.

REFERENCES

1. Barudov S., Barudov E., Discrete alternating current regulators and stabilizers, PENSOFT Sofia-Moscow 2006.

2. Harlow James H., Transformers. The electric power engineering handbook, Ed. L. L. Grigsby Boca Raton: CRC Press LLC, 2001.

3. Barudov E., Barudov S., Panov E., Switching processes in a step voltage regulator, Acta Universitatis Pontica Euxinus, Volume 4, Number 1, 2005, pp. 21÷25.

4. Barudov S., Panov E., Study of the loading of the switching elements in a step voltage regulator, Annual Proceedings of Technical University of Varna, Varna, 2004, p. 117÷122.

5. Barudov E., Panov E., Barudov S., Study of a precise non-linear model of an autotransformer discrete voltage regulator with semiconductor switching elements, Annual of TU-Varna, 2007, pp. 91÷96.

6. Barudov S., Barudov E. Step-up regulators of variable voltage, SPb: PEIPK, Methods and tools for assessing the condition of power equipment. Vol. 25, 2005, pp. 217-222.

7. Panov E., Barudov E., Barudov S., Advanced algorithm for the analysis of a precise non-linear model of an autotransformer discrete voltage regulator with semiconductor switching elements, Annual of TU-Varna, 2009, pp. 47-52, ISSN: 1311-896X.

8. Barudov E., Barudov S., Panov E., Study of the transient process length in step voltage regulator, Acta Universitatis Pontica Euxinus, Volume 3, Number 1, 2004, p. 91÷96.

9. Barudov S., Panov E., Barudov E., Parameters of the commutation process in the step regulator for exchange voltage. Proceedings of the Petersburg Power Engineering Institute of Advanced Training SPb: PEIPK, Methods and Tools for Estimation of Power Engineering Equipment, Vol. 25, 2005, pp. 67÷76, UDK 621.3.048, 621.315.62, BBK 31.264-04.

10. Chua L., Lin P., Machine analysis of electronic circuits (algorithms and computational methods), Moscow, Energia, 1980.

11. Hambley A., Top-Down Approach to Computer – Aided Circuit Design, Prentice Hall, Englewood Chiffs, New Jersey, 1994.

12. Bose B. K. Modern power electronics: Evaluation, technology and applications, IEEE New York, 1992.

13. Comer D., Computer Analysis of Circuits, International Textbook Company, 1971.

ANALYSIS OF ELECTRICAL PROCESSES IN A DISCRETE AC VOLTAGE REGULATOR UNDER RESISTIVE-CAPACITIVE LOAD

Emil Barudov, Emil Panov, Stefan Barudov

(Published in the Proceedings of the International Scientific and Technical Conference "Electrical Power Engineering 2010", 14-16 October 2010, Varna, pp. 332-341, ISBN 978-954-20-0497-4. (in Bulgarian))

Abstract: **One of the problems related to power quality is the regulation of supply voltage amplitude. The work is devoted to the problems of the analysis of autotransformer discrete alternating voltage regulators (ADAVR) and includes simulation models considering the specifics of the chosen circuit solution. A specialized algorithm is developed for solving piecewise the resulting sparse matrix equations with cell structure, which have almost singular matrices. The resulting solutions have little error accumulation from the computations and are highly stable in time. Experimental verification with a prototype has been performed and shows good coincidence with the simulated process data.**

Keywords: **discrete ac voltage regulator, semiconductor commutation elements, thyristor, resistive-capacitive load**

1. INTRODUCTION

Recently, there has been a growing interest in more accurately controlling the quality of electricity supplied for industrial and domestic use [1]. One of the important elements is the amplitude variation range of the input supply voltage. The control of the supply voltage can be realized by using autotransformer discrete alternating voltage regulators (ADAVR) with semiconductor commutation elements (SCE) [2, 3]. The design and construction of ADAVR with SCE requires the use of advanced tools for computer analysis and simulation of the complex processes occurring in these devices [4, 5, 6, 7, 8]. The main problem is the development of adequate accurate models and analysis algorithms that allow accurate simulations of ADAVR responses under different operating modes, which leads to the realization of savings in money and time in the design and fabrication of such devices.

2. ANALYSIS

The limitation of the amplitude variation range of the input supply voltage can be realized by the automatic switching of the transformer terminals (autotransformer). The terminals can be located on the mains side or on the load side [1]. Fig. 1 shows the replacement scheme of the ADAVR with four SCE. The scheme includes the magnetic circuit parameters, the switching groups, the switch-off groups, the parameters of the individual winding sections and takes into account the existing non-linearities.

The specificity of thyristor commutators is taken into account, with one switching process covering three intervals. In the simulations, the most severe mode for the ADAVR is investigated, i.e. in the first interval the K_3 switch is closed and all others are open; in the second interval the K_3 and K_4 switches are closed; in the third interval only the K_4 switch is closed. The analysis carried out is for resistive-capacitive loads, and a detailed algorithm is presented for solving the system of equations describing the electrical equilibrium in the ADAVR.

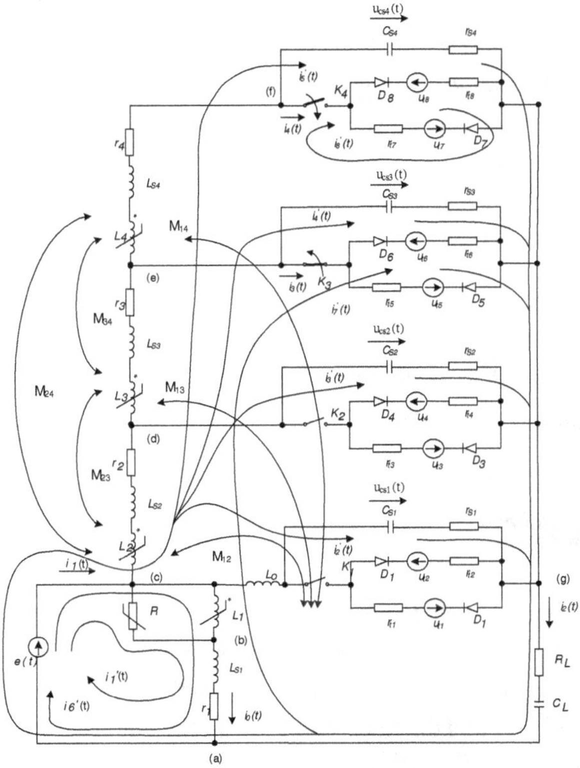

Fig. 1. Replacement scheme of ADAVR with four SCE and resistive-capacitive load.

Fig. 2 shows the graph of the studied circuit for the second interval of the switching process.

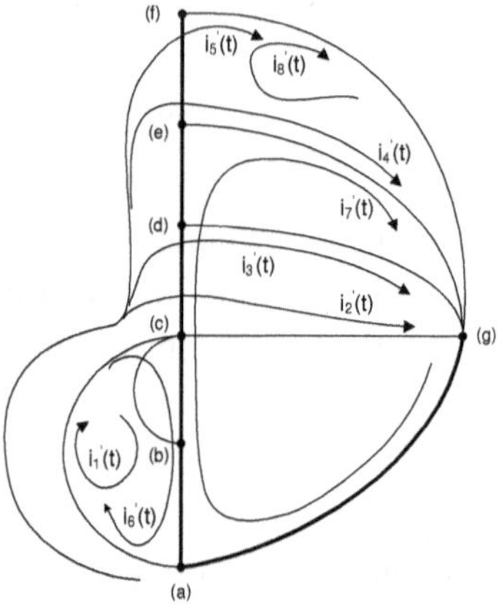

Fig. 2. Graph of the circuit for the II interval.

In the first interval, the ADAVR operates in steady-state AC mode, with the electrical equilibrium described by means of the loop current method in complex form. The number of independent loops in the circuit is seven and the system of equations consists of seven equations with seven unknowns. The general form of the system has the following matrix form:

$$[Y_1].[\dot{I}'] = [\dot{E}]\,, \tag{1}$$

where $[Y_1]$ is a complex conductivity matrix of ADAVR with dimensionality 7 rows and 7 columns (i.e. 7x7); $[\dot{I}']$ is a matrix-column of the phasors of the loop currents $\dot{I}_1'$; $\dot{I}_2'$;...$\dot{I}_7'$ with dimensionality 7x1; $[\dot{E}]$ is a matrix-column of the phasors of the loop electromotive forces with a dimensionality 7x1. The numeric method used to solve equation (1) is the Gauss-Jordan method.

In the second interval, the first transition process is observed in the ADAVR - switches K3 and K4 are closed. The analysis is performed using the state variable approach, where eight of the state variables are the loop currents $i_1'(t);i_2'(t);...;i_7'(t);i_8'(t)$, and the remaining five are the voltage drops across the load capacitance $u_{CL}(t)$ and the voltage drops over the switch-off capacitors $u_{cs1}(t);...;u_{cs4}(t)$. The analysis is conducted in the time domain as the system of equations contains thirteen equations with thirteen unknowns. The matrix of the system of equations appears almost singular regardless of the type of loops chosen. And this is a prerequisite for the accumulation of error in the calculations. In this situation, the general matrix form of the system of equations is as follows:

$$[A_2]\frac{d}{dt}[x(t)] = [B_2].[x(t)] + [e_2(t)] , \qquad (2)$$

where the matrices $[A_2]$ and $[B_2]$ are with a dimensionality 13x13 and they consist of nine sub-matrices:

$$[A_2] = \left[\begin{array}{c|c|c} A_{11} & A_{12} & A_{13} \\ \hline A_{21} & A_{22} & A_{23} \\ \hline A_{31} & A_{32} & A_{33} \end{array}\right], \quad [B_2] = \left[\begin{array}{c|c|c} B_{11} & B_{12} & B_{13} \\ \hline B_{21} & B_{22} & B_{23} \\ \hline B_{31} & B_{32} & B_{33} \end{array}\right]. \qquad (3)$$

The matrices $[x(t)]$ and $[e_2(t)]$ are with dimensionality 13x1 and they consist of three matrices each:

$$[x(t)] = \begin{bmatrix} X_{11}(t) \\ \overline{X_{21}(t)} \\ \overline{X_{31}(t)} \end{bmatrix}, \quad [e_2(t)] = \begin{bmatrix} E_{11}(t) \\ \overline{E_{21}(t)} \\ \overline{E_{31}(t)} \end{bmatrix}. \qquad (4)$$

Here, the matrix $[x(t)]$ contains the state variables in the order listed above, and the matrix $[e_2(t)]$ contains the input electromotive force e(t) and the voltage drops over the thyristor commutators K_3 and K_4 - u_{t5}, u_{t6}, u_{t7} and u_{t8} in saturated state.

The sizes of the submatrices are as follows:

$[A_{11}]$ and $[B_{11}]$ are 5x5; $[A_{12}]$ and $[B_{12}]$ are 5x4; $[A_{13}]$ and $[B_{13}]$ are 5x4;

$[A_{21}]$ and $[B_{21}]$ are 4x5; $[A_{22}]$ and $[B_{22}]$ are 4x4; $[A_{23}]$ and $[B_{23}]$ are 4x4;

$[A_{31}]$ and $[B_{31}]$ are 4x5; $[A_{32}]$ and $[B_{32}]$ are 4x4; $[A_{33}]$ and $[B_{33}]$ are 4x4;

$[X_{11}(t)]$ and $[E_{11}(t)]$ are 5x1; $[X_{21}(t)]$ and $[E_{21}(t)]$ are 4x1; $[X_{31}(t)]$ and $[E_{31}(t)]$ are 4x1.

The matrices $[A_2]$ and $[B_2]$ are sparse, in which the following submatrices are null: $[A_{13}]$; $[A_{23}]$; $[A_{31}]$; $[A_{32}]$; $[B_{23}]$; $[B_{31}(:,1)]$; $[B_{32}]$; $[B_{33}]$; $[E_{31}(t)]$. Except that, the submatrices $[A_{33}]$, $[B_{13}]$, $[B_{31}]$ are diagonal. The submatrices of the state variables have the following form:

$$[X_{11}(t)] = \begin{bmatrix} i_1^{'}(t) \\ i_2^{'}(t) \\ \vdots \\ i_5^{'}(t) \end{bmatrix} ; \ [X_{21}(t)] = \begin{bmatrix} i_6^{'}(t) \\ i_7^{'}(t) \\ i_8^{'}(t) \\ --- \\ u_{CT}(t) \end{bmatrix} ; \ [X_{31}(t)] = \begin{bmatrix} u_{cs1}(t) \\ u_{cs2}(t) \\ u_{cs3}(t) \\ u_{cs4}(t) \end{bmatrix} . \tag{5}$$

The algorithm, which is applied to avoid the fact that the matrix $[A_2]$ is almost singular, consists in breaking equation (2) into three matrix equations containing the above listed submatrices, and the solutions are obtained in a sequential order. At each time step forward, the following matrix equations are solved:

$$1) \ [A_{169}]\frac{d}{dt}\begin{bmatrix} i_1^{'}(t) \\ ---- \\ X_{21}(t) \end{bmatrix} = [B_{169}].\begin{bmatrix} i_1^{'}(t) \\ ---- \\ X_{21}(t) \end{bmatrix} + \begin{bmatrix} e_2(t) \\ ---- \\ E_{21}(t) \end{bmatrix} , \tag{6}$$

where the current $i_1^{'}(t)$ is the loop current through Rμ (the resistance accounting for losses in the autotransformer steel), and the matrices $[A_{169}]$ and $[B_{169}]$ contain respectively the submatrices $[A_{22}]$ and $[B_{22}]$ plus the elements connected with the current $i_1^{'}(t)$ from equation (2). The solution of the matrix differential equation (6) is performed after it is normalized by left and right multiplication using the inverse matrix $[A_{169}]^{-1}$ and for the received equation:

$$\begin{aligned} \frac{d}{dt}\begin{bmatrix} i_1^{'}(t) \\ ------ \\ X_{21}(t) \end{bmatrix} &= [A_{169}]^{-1} . \ [B_{169}].\begin{bmatrix} i_1^{'}(t) \\ ------ \\ X_{21}(t) \end{bmatrix} + [A_{169}]^{-1} . \begin{bmatrix} e_2(t) \\ ------ \\ E_{21}(t) \end{bmatrix} = \\ &= [AA_2] . \begin{bmatrix} i_1^{'}(t) \\ ------ \\ X_{21}(t) \end{bmatrix} + [BB_2] . \begin{bmatrix} e_2(t) \\ ------ \\ E_{21}(t) \end{bmatrix} \end{aligned} \tag{7}$$

the Cauchy's formula for exact solution of systems of inhomogeneous differential equations is applied:

$$\begin{bmatrix} i_1^{'}(t) \\ ------ \\ X_{21}(t) \end{bmatrix} = e^{[AA_2].(t-t_0)} . \begin{bmatrix} i_1^{'}(0+) \\ ------ \\ X_{21}(0+) \end{bmatrix} + e^{[AA_2].t} . \int_{t_0}^{t} e^{-[AA_2].\tau} . [B_2] . \begin{bmatrix} e_2(\tau) \\ ------ \\ E_{21}(\tau) \end{bmatrix} . d\tau \tag{8}$$

2) $[A_{33}]\frac{d}{dt}[X_{32}(t)]=[B_{31}(:,2:5)]\,.\,[X_{11}(t)]$, (9)

as equation (9) is normalized by multiplying left and right by the inverse matrix $[A_{33}]^{-1}$ and then solved using the Runge-Kutta method – 4.

3)
$$[A_{11}(2:5,2:5)]\,.\frac{d}{dt}[X_{11}(:,2:5)]+\begin{bmatrix}A_{11}(2:5,1)\\ \text{---------}\\ A_{12}(2:5,:)\end{bmatrix}.\frac{d}{dt}\begin{bmatrix}i_1'(t)\\ \text{------}\\ X_{21}(t)\end{bmatrix}= \tag{10}$$
$$=\left[B_{11}(2:5,:)\middle|B_{12}(2:5,:)\middle|B_{13}(2:5,:)\right]\,.\,[x(2:13,1)]+[e_2(2:13,1)]\,,$$

as equation (10) is normalized and solved using the Runge-Kutta method – 4.

In the third interval, the second transition process is observed, with only K_4 closed of all the switches. The analysis is again conducted using the time domain state variable approach. The number of state variables is now twelve, since the loop current $i_7'(t)$ from the group of unknowns from the previous interval is dropped here. The system of equations in the third interval has the following form:

$$[A_3]\frac{d}{dt}[x(t)]=[B_3].[x(t)]+[e_3(t)]\,, \tag{11}$$

where the matrices $[A_3]$ are $[B_3]$ are with dimensionality 12x12 and they consist of three sub-matrices each:

$$[A_3]=\begin{bmatrix}a_{11}| & a_{12}| & a_{13}\\ \hline a_{21}| & a_{22}| & a_{23}\\ \hline a_{31}| & a_{32}| & a_{33}\end{bmatrix},\quad [B_3]=\begin{bmatrix}b_{11}| & b_{12}| & b_{13}\\ \hline b_{21}| & b_{22}| & b_{23}\\ \hline b_{31}| & b_{32}| & b_{33}\end{bmatrix}. \tag{12}$$

The matrices $[x(t)]$ and $[e_3(t)]$ are with dimensionality 12x1 and they consist of three sub-matrices each:

$$[x(t)]=\begin{bmatrix}x_{11}(t)\\ \overline{x_{21}(t)}\\ \overline{x_{31}(t)}\end{bmatrix},\quad [e_3(t)]=\begin{bmatrix}e_{11}(t)\\ \overline{e_{21}(t)}\\ \overline{e_{31}(t)}\end{bmatrix}. \tag{13}$$

The matrix $[x(t)]$ contains the state variables in the order listed for the second interval without the current $i_7'(t)$, and the matrix $[e_3(t)]$ contains the input electromotive force e(t) and the voltage drops over the thyristor commutator K_4 - u_{t7} and u_{t8} in saturated state.

The dimensionality of the matrices are as follows:

$[a_{11}]$ and $[b_{11}]$ are 5x5; $[a_{12}]$ and $[b_{12}]$ are 5x3; $[a_{13}]$ and $[b_{13}]$ are 5x4;

$[a_{21}]$ and $[b_{21}]$ are 3x5; $[a_{22}]$ and $[b_{22}]$ are 3x3; $[a_{23}]$ and $[b_{23}]$ are 3x4;

$[a_{31}]$ and $[b_{31}]$ are 4x5; $[a_{32}]$ and $[b_{32}]$ are 4x3; $[a_{33}]$ and $[b_{33}]$ are 4x4;

$[x_{11}(t)]$ and $[e_{11}(t)]$ are 5x1; $[x_{21}(t)]$ and $[e_{21}(t)]$ are 3x1; $[x_{31}(t)]$ and $[e_{31}(t)]$ are 4x1.

The matrices $[A_3]$ and $[B_3]$ are sparse, as the matrix $[A_3]$ is almost singular. The following submatrices are also null: $[a_{13}]$; $[a_{23}]$; $[a_{31}]$; $[a_{32}]$; $[b_{23}]$; $[b_{32}]$; $[b_{33}]$; $[e_{31}(t)]$. Except that, the submatrices $[a_{33}]$, $[b_{13}]$, $[b_{31}]$ are diagonal. The submatrices of the state variables have the following form:

$$[x_{11}(t)]=\begin{bmatrix} i_1^{'}(t) \\ i_2^{'}(t) \\ \vdots \\ i_5^{'}(t) \end{bmatrix} \quad [x_{21}(t)]=\begin{bmatrix} i_6^{'}(t) \\ i_8^{'}(t) \\ --- \\ u_{CT}(t) \end{bmatrix}; \quad [x_{31}(t)]=\begin{bmatrix} u_{cs1}(t) \\ u_{cs2}(t) \\ u_{cs3}(t) \\ u_{cs4}(t) \end{bmatrix}. \tag{14}$$

The algorithm, which avoids the fact that the matrix $[A_3]$ is almost singular, consists in breaking equation (11) into three matrix equations containing the above sub-matrices, performing at each time step forward a similar computational procedure as in the second interval described by equations (6), (7), (8), (9) and (10).

In Fig. 3, the plots of the investigated quantities in the three intervals of the switching process of the ADAVR with resistive-capacitive load are presented (R_L=115,8Ω and C_L=38,09μF). The phase of commutation of the switch K_4 is chosen to be φ=225°. In Fig. 3a the input current $i_1(t)$ is presented as a function of time; in Fig. 3b the output current $i_2(t)$ is shown; in Fig. 3c – the current $i_4(t)$ through the switch K_4; in Fig. 3d – the voltage drop $u_{cs3+rs3}(t)$ over the third switch-off group $C_{S3}+r_{S3}$; in Fig. 3e – the current $i_{cs4+rs4}(t)$ through the fourth switch-off group and in Fig. 3f – the voltage drop $u_{cs4+rs4}(t)$ over the fourth switched-off group $C_{S4}+r_{S4}$.

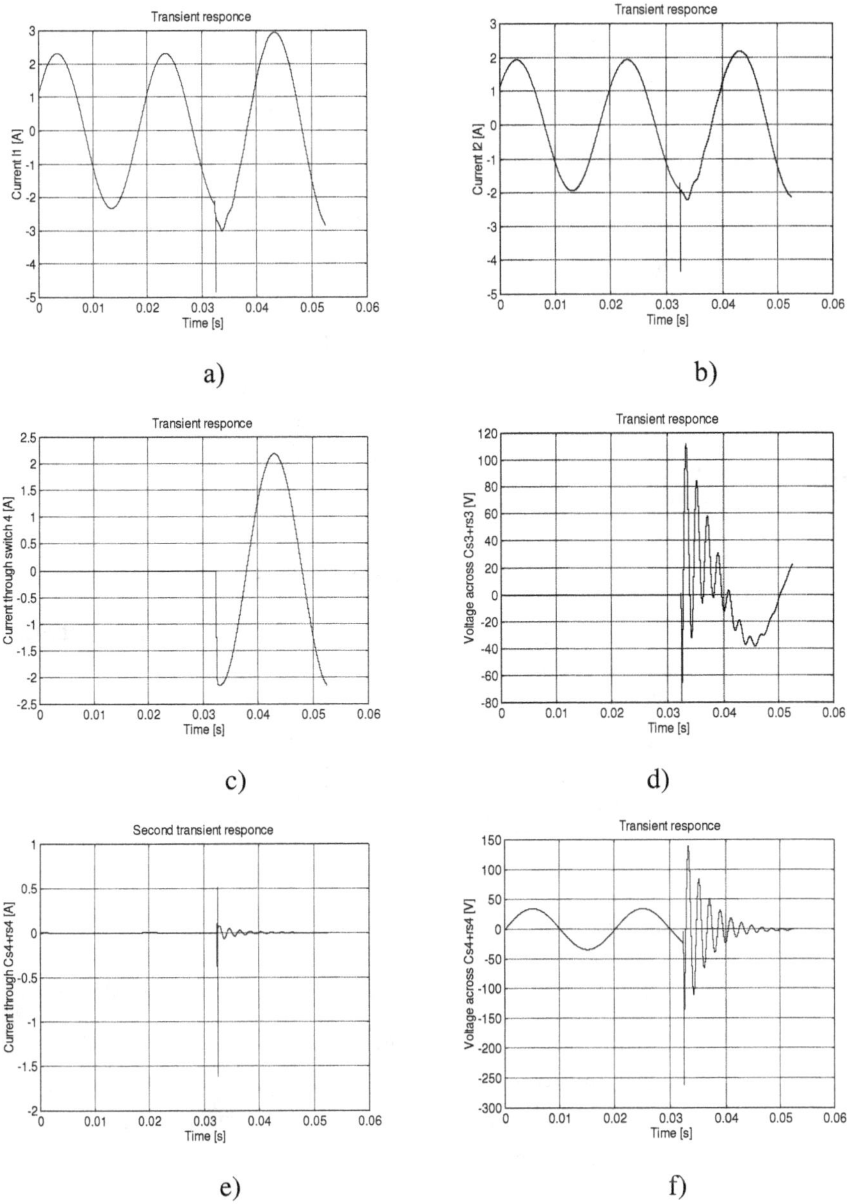

Fig. 3. Plots of the investigated quantities in ADAVR with resistive-capacitive load for the three intervals of the commutation process.

Table 1a and 1b present the experimental data and the computer simulations using the proposed algorithm and the program AVTO for a particular ADAVR at an input voltage of 160V with three different resistive and two different capacitive loads.

Table 1. Results of physical experiments and computer simulations.

a) C_L=9,14μF

	Results of the physical experiments						*Results of the computer simulations*					
R_L	U_1	I_1	I_0	I_2	U_C	U_2	U_1	I_1	I_0	I_2	U_C	U_2
Ω	V	A	A	A	V	V	V	A	A	A	V	V
115,8	160	0,7	0,13	0,62	208,6	220	160	0,693	0,152	0,5953	207,3	218,5
61,8	160	0,73	0,15	0,64	216,3	220,6	160	0,710	0,133	0,6224	216,8	220,16
35,5	160	0,73	0,14	0,66	219	220,8	160	0,717	0,122	0,6304	219,6	220,7

б) C_L=38,09μF

	Results of the physical experiments						*Results of the computer simulations*					
R_L	U_1	I_1	I_0	I_2	U_C	U_2	U_1	I_1	I_0	I_2	U_C	U_2
Ω	V	A	A	A	V	V	V	A	A	A	V	V
115,8	160	2,1	0,61	1,56	130,2	220,3	160	2,088	0,568	1,5442	129,0	220,50
61,8	160	2,85	0,75	2,15	174,9	220,3	160	2,885	0,723	2,1700	181,3	225,54
35,5	160	3,3	0,85	2,6	205,8	220,7	160	3,253	0,802	2,4628	205,8	223,66

3. CONCLUSIONS

From the presented research the following conclusions can be drawn:

1. A detailed algorithm for solving the system of equations describing the electrical equilibrium in the ADAVR under resistive-capacitive load is proposed, which allows overcoming the peculiarity of the system matrix in the second and third interval of the switching process of the voltage regulator.

2. On the basis of the proposed algorithm, the AVTO computer program is further developed, covering the specificity of the circuit solution, the variation of the autotransformer parameters as a function of the mode, the magnitude and nature of the load, the phase of commutations and the parameters of the semiconductor commutators used. This enables the nature and duration of the switching process, as well as the electrical load on the individual circuits and semiconductor switches, to be quantified.

3. The comparison between the data of computer simulations and physical experiments for the investigated switches is presented in tabular form, which shows a good coincidence of the obtained results.

REFERENCES:

1. Barudov S., Barudov E., Discrete alternating current regulators and stabilizers, PENSOFT Sofia-Moscow 2006.

2. Barudov E., Barudov S., Panov E., Switching processes in a step voltage regulator, Acta Universitatis Pontica Euxinus, Volume 4, Number 1, 2005, pp. 21÷25.

3. Harlow James H., Transformers. The electric power engineering handbook, Ed. L. L. Grigsby Boca Raton: CRC Press LLC, 2001.

4. Barudov S., Panov E., Study of the loading of the switching elements in a step voltage regulator, Annual Proceedings of Technical University of Varna, Varna, 2004, pp. 117÷122.

5. Barudov E., Panov E., Barudov S., Study of a precise non-linear model of an autotransformer discrete voltage regulator with semiconductor commutation elements, Annual of TU-Varna, 2007, pp. 91÷96.

6. Panov E., Barudov E., Barudov S., Advanced algorithm for analysis of a precise non-linear model of an autotransformer discrete voltage regulator with semiconductor commutation elements, Annual of TU-Varna, 2009, Bulgaria, pp. 47÷52, ISSN: 1311-896X.

7. Barudov E., Barudov S., Panov E., Study of the transient process length in step voltage regulator, Acta Universitatis Pontica Euxinus, Volume 3, Number 1, 2004, pp. 91÷96, ISSN 1312-1669.

8. Barudov E., Panov E., Barudov S., Analysis of Electrical Processes in Alternating Voltage Control Systems, 12-th International Symposium "Materials, Methods&Technologies (MMT), 11-15 June 2010, Sunny Beach, Bulgaria, published in Journal of International Scientific Publications: Materials, Methods & Technologies, Volume 4, Part 1, 2010, pp. 154 – 182, ISSN: 1313 2539, http://www.science-journals.eu.

ADDRESSES OF THE AUTHORS

1. Emil Barudov, Technical University - Varna, 9010, Bulgaria, 1 Studentska str., Faculty of Electrical Engineering, Department TIE, e-mail: ugl@gyuvetch.bg

2. Emil Panov, Technical University - Varna, 9010, Bulgaria, 1 Studentska str., Faculty of Electrical Engineering, Department TIE, e-mail: eipanov@yahoo.com

3. Stefan Barudov, Technical University - Varna, 9010, Bulgaria, 1 Studentska str., Faculty of Electrical Engineering, Department EE, e-mail: sbarudov@abv.bg

ANALYSIS OF ELECTRICAL PROCESSES IN A DISCRETE ALTERNATING VOLTAGE REGULATOR WITH ACTIVE-INDUCTIVE LOAD

Emil Barudov, Emil Panov, Stefan Barudov

(Published in the Annual of the Technical University - Varna, Vol. II, 2010, pp. 30 - 35, ISSN: 1311-896X. (in Bulgarian)

Abstract: In networks with limited power, the voltage supply of the separate consumers can be controlled in admissible range through autotransformer discrete alternating voltage regulators. This paper presents the problems of the analysis, through simulation models while accounting for the specific features of a particular circuit design. A special algorithm has been developed for part-by-part solving of the obtained sparse matrix equations with cell structure, which have almost singular matrices. The obtained solutions show insignificant error accumulation from the computations and great stability in time. The data from the simulated processes have been experimentally verified using a prototype and have shown a good correspondence.

Keywords: autotransformer, discrete alternating voltage regulator, semiconductor commuting elements, thyristor, commutation process, active-inductive load

I. INTRODUCTION

There are electrical networks for which, depending on the operating modes for some of the end consumers, the phase voltage varies far beyond the permissible limits. This necessitates the regulation of the input supply voltage variation range, which can be realized by using autotransformer discrete alternating voltage regulators (ADAVR) with semiconductor commutation elements (SCE) [1, 2, 3]. The design of such regulators requires the use of tools for computer analysis and simulation of the complex processes occurring in these devices [3, 4, 5, 6, 7]. The main problem is to create adequate precise models and analysis algorithms that allow accurate simulations of the responses of the ADAVR under different operating modes. This leads to savings of materials, elements and time in the design and the fabrication phase of such a class of devices.

II. ANALYSIS

The limitation of the amplitude variation range of the input supply voltage is realized by the automatic commutation of the transformer terminals (autotransformer). Circuit solutions allow the autotransformer terminals to be located on the mains side or on the load side [1, 2]. In Fig. 1, the replacement scheme of the ADAVR with four SCE is presented. The scheme includes the parameters of the commutation groups, the switch-off groups, the magnetic circuit, the parameters of the individual winding sections and the existing non-linearities.

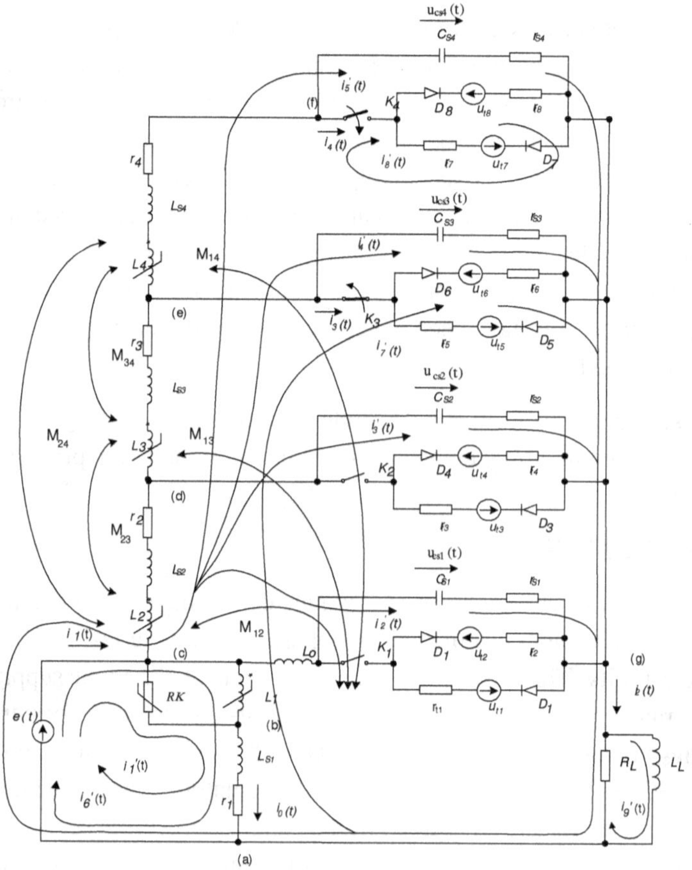

Fig. 1. Replacement scheme of ADAVR with four SCE and active-inductive load.

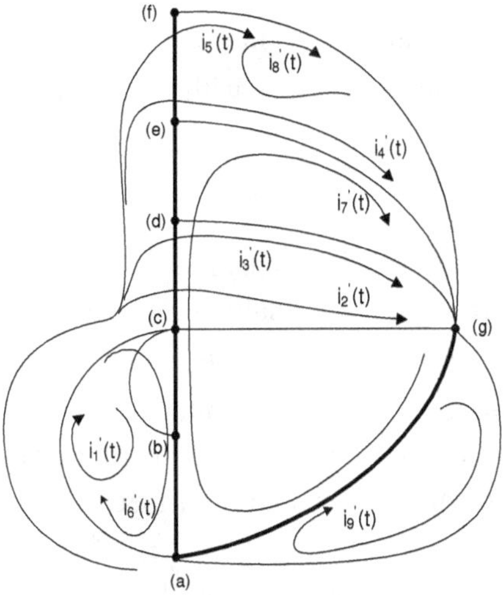

Fig. 2. Graph of the circuit for the II interval.

In the simulations, the most severe mode of operation was investigated - switching from the penultimate to the last increasing degree of ADAVR. The commutation process covers three intervals [3, 4]. In the first interval, switch K_3 is closed and all others are open; in the second interval, switches K_3 and K_4 are closed; and in the third interval, only switch K_4 is closed.

In this work the analysis for active-inductive load is performed and a detailed algorithm for solving the system of equations describing the electrical equilibrium in the ADAVR is presented.

Fig. 2 shows the graph of the investigated circuit for the second interval of the commutation process.

In the first interval, the ADAVR operates at steady-state regime and the electrical equilibrium is described by the loop current method in complex form.

The number of independent loops in the circuit is eight and the system of equations consists of eight equations with eight unknowns. The general form of the system has the following matrix form:

$$[Y_1].[\dot{I}'] = [\dot{E}], \tag{1}$$

where $[Y_1]$ is a complex conductivity matrix of ADAVR with dimension 8 rows and 8 columns (i.e. 8x8); $[\dot{I}']$ is a matrix-column of the phasors of the loop currents $\dot{I}_1'; \dot{I}_2';....;\dot{I}_7';\dot{I}_9'$ with dimension 8x1; $[\dot{E}]$ is a matrix-column of the phasors of the loop electromotive forces with a dimension 8x1.

The numerical method, by which equation (1) is solved, is the Gauss-Jordan method.

In the second interval, the first transition process is observed in the ADAVR - the keys K3 and K4 are closed. The analysis is performed using the state variable approach, nine of the state variables are the loop currents $i_1'(t);i_2'(t);...;i_8'(t);i_9'(t)$, and the remaining four variables are the voltages upon the switch-off capacitors $u_{cs1}(t);\ldots;u_{cs4}(t)$. The analysis is conducted in the time domain, and the system of equations contains thirteen equations with thirteen unknowns. The matrix of the system of equations appears almost peculiar, regardless of the type of the loops chosen, resulting in an accumulated error in the calculations. Then, the general matrix form of the system of equations is as follows:

$$[A_2]\frac{d}{dt}[x(t)]=[B_2].[x(t)]+[e_2(t)], \qquad (2)$$

where the matrices $[A_2]$ and $[B_2]$ are with dimensions 13x13 and they consist of nine submatrices each:

$$[A_2]=\left[\begin{array}{c|c|c} A_{11} & A_{12} & A_{13} \\ \hline A_{21} & A_{22} & A_{23} \\ \hline A_{31} & A_{32} & A_{33} \end{array}\right],[B_2]=\left[\begin{array}{c|c|c} B_{11} & B_{12} & B_{13} \\ \hline B_{21} & B_{22} & B_{23} \\ \hline B_{31} & B_{32} & B_{33} \end{array}\right]. \qquad (3)$$

The matrices $[x(t)]$ and $[e_2(t)]$ are with dimensions 13x1 and they consist of three submatrices each:

$$[x(t)]=\begin{bmatrix} X_{11}(t) \\ \overline{X_{21}(t)} \\ \overline{X_{31}(t)} \end{bmatrix}, \quad [e_2(t)]=\begin{bmatrix} E_{11}(t) \\ \overline{E_{21}(t)} \\ \overline{E_{31}(t)} \end{bmatrix}. \qquad (4)$$

Here, the matrix $[x(t)]$ contains the state variables in the order listed above, and the $[e_2(t)]$ contains the input electromotive force e(t) and the voltage drops over the thyristor commutators K_3 and K_4 - u_{t5}, u_{t6}, u_{t7} and u_{t8} in saturated state.

The dimensions of the submatrices are as follows:

$[A_{11}]$ and $[B_{11}]$ are 5x5; $[A_{12}]$ and $[B_{12}]$ are 5x4; $[A_{13}]$ and $[B_{13}]$ are 5x4;

$[A_{21}]$ and $[B_{21}]$ are 4x5; $[A_{22}]$ and $[B_{22}]$ are 4x4; $[A_{23}]$ and $[B_{23}]$ are 4x4;

$[A_{31}]$ and $[B_{31}]$ are 4x5; $[A_{32}]$ and $[B_{32}]$ are 4x4; $[A_{33}]$ and $[B_{33}]$ are 4x4;

$[X_{11}(t)]$ and $[E_{11}(t)]$ are 5x1; $[X_{21}(t)]$ and $[E_{21}(t)]$ are 4x1; $[X_{31}(t)]$ and $[E_{31}(t)]$ are 4x1.

The matrices $[A_2]$ and $[B_2]$ are sparse, with the following submatrices appearing as null: $[A_{13}]$; $[A_{23}]$; $[A_{31}]$; $[A_{32}]$; $[B_{23}]$; $[B_{31}(:,1)]$; $[B_{32}]$; $[B_{33}]$; $[E_{31}(t)]$. Except that, the submatrices $[A_{33}]$, $[B_{13}]$, $[B_{31}]$ are diagonal. The submatrices of the state variables are of the following type:

$$[X_{11}(t)]=\begin{bmatrix} i_1'(t) \\ i_2'(t) \\ \vdots \\ i_5'(t) \end{bmatrix};[X_{21}(t)]=\begin{bmatrix} i_6'(t) \\ i_7'(t) \\ i_8'(t) \\ i_9'(t) \end{bmatrix};[X_{31}(t)]=\begin{bmatrix} u_{cs1}(t) \\ u_{cs2}(t) \\ u_{cs3}(t) \\ u_{cs4}(t) \end{bmatrix}. \quad (5)$$

The algorithm that is applied, to avoid the fact that the matrix $[A_2]$ is almost singular (degenerate), consists in breaking equation (2) into three matrix equations containing the above submatrices, with the solutions being obtained in sequential order. At each time step forward, the following matrix equations are solved:

$$1) \quad [A_{169}]\frac{d}{dt}\begin{bmatrix} i_1'(t) \\ ---- \\ X_{21}(t) \end{bmatrix}=[B_{169}].\begin{bmatrix} i_1'(t) \\ ---- \\ X_{21}(t) \end{bmatrix}+\begin{bmatrix} e_2(t) \\ ---- \\ E_{21}(t) \end{bmatrix}, \quad (6)$$

where the current $i_1'(t)$ is the loop current through $R\mu$ (the resistance accounting for losses in the autotransformer steel core), and the matrices $[A_{169}]$ and $[B_{169}]$ contain the submatrices $[A_{22}]$ and $[B_{22}]$, respectively, plus the elements related with the current $i_1'(t)$ from equation (2). The solution of the matrix differential equation (6) is performed after it is normalized by left and right multiplication using the inverse matrix $[A_{169}]^{-1}$ and for the resulting equation:

$$\frac{d}{dt}\begin{bmatrix} i_1'(t) \\ ----- \\ X_{21}(t) \end{bmatrix}=[A_{169}]^{-1}.\,[B_{169}].\begin{bmatrix} i_1'(t) \\ ----- \\ X_{21}(t) \end{bmatrix}+[A_{169}]^{-1}.\begin{bmatrix} e_2(t) \\ ----- \\ E_{21}(t) \end{bmatrix}=$$
$$=[AA_2]\,.\begin{bmatrix} i_1'(t) \\ ----- \\ X_{21}(t) \end{bmatrix}+[BB_2]\,.\begin{bmatrix} e_2(t) \\ ----- \\ E_{21}(t) \end{bmatrix} \quad (7)$$

The Cauchy formula is applied to exactly solve systems of inhomogeneous differential equations:

$$\begin{bmatrix} i_1'(t) \\ ---- \\ X_{21}(t) \end{bmatrix} = e^{[AA2].(t-t_0)} . \begin{bmatrix} i_1'(0+) \\ ---- \\ X_{21}(0+) \end{bmatrix} + e^{[AA2].t} . \int_{t_0}^{t} e^{-[AA2].\tau} . [B_2] . \begin{bmatrix} e_2(\tau) \\ ---- \\ E_{21}(\tau) \end{bmatrix} .d\tau \quad (8)$$

2) $$[A_{33}]\frac{d}{dt}[X_{32}(t)] = [B_{31}(:,2:5)].[X_{11}(t)] , \quad (9)$$

as equation (9) is normalized by multiplying left and right by the inverse matrix $[A_{33}]^{-1}$ and then solved using the Runge-Kutta method - 4.

3) $$[A_{11}(2:5,2:5)] .\frac{d}{dt}[X_{11}(:,2:5)] + \begin{bmatrix} A_{11}(2:5,1) \\ --------- \\ A_{12}(2:5, :) \end{bmatrix} .\frac{d}{dt}\begin{bmatrix} i_1'(t) \\ ----- \\ X_{21}(t) \end{bmatrix} = \quad (10)$$

$$= \left[B_{11}(2:5, :) | B_{12}(2:5, :) | B_{13}(2:5, :) \right] . [x(2:13, 1)] + [e_2(2:13, 1)] ,$$

as equation (10) is normalized and solved using the Runge-Kutta method - 4.

In the third interval, the second transition process is observed, and of all the switches, only the K4 switch remains closed. The analysis is again conducted using the time domain state variable method. The number of state variables is now twelve, since the loop current $i_7'(t)$ from the group of unknowns from the previous interval is dropped here.

The system of equations in the third interval has the following form:

$$[A_3]\frac{d}{dt}[x(t)] = [B_3].[x(t)] + [e_3(t)] , \quad (11)$$

where the matrices $[A_3]$ and $[B_3]$ are with a dimension 12x12 and they consist of three submatrices each:

$$[A_3] = \begin{bmatrix} a_{11} | & a_{12} | & a_{13} \\ \hline a_{21} | & a_{22} | & a_{23} \\ \hline a_{31} | & a_{32} | & a_{33} \end{bmatrix} , \quad [B_3] = \begin{bmatrix} b_{11} | & b_{12} | & b_{13} \\ \hline b_{21} | & b_{22} | & b_{23} \\ \hline b_{31} | & b_{32} | & b_{33} \end{bmatrix} . \quad (12)$$

The matrices $[x(t)]$ and $[e_3(t)]$ are with dimensions 12x1 and they consist of three submatrices each:

$$[x(t)] = \begin{bmatrix} x_{11}(t) \\ \overline{x_{21}(t)} \\ \overline{x_{31}(t)} \end{bmatrix} , \quad [e_3(t)] = \begin{bmatrix} e_{11}(t) \\ \overline{e_{21}(t)} \\ \overline{e_{31}(t)} \end{bmatrix} . \quad (13)$$

The matrix [x(t)] contains the state variables in the order listed for the second interval without the current $i'_7(t)$, and the matrix [$e_3(t)$] contains the input electromotive force e(t) and the voltage drops upon the switch K_4 - u_{t7} and u_{t8} in saturated state.

The dimensions of the submatrices are as follows:

[a_{11}] and [b_{11}] are 5x5; [a_{12}] and [b_{12}] are 5x3; [a_{13}] and [b_{13}] are 5x4;

[a_{21}] and [b_{21}] are 3x5; [a_{22}] and [b_{22}] are 3x3; [a_{23}] and [b_{23}] are 3x4;

[a_{31}] and [b_{31}] are 4x5; [a_{32}] and [b_{32}] are 4x3; [a_{33}] and [b_{33}] are 4x4;

[$x_{11}(t)$] and [$e_{11}(t)$] are 5x1; [$x_{21}(t)$] and [$e_{21}(t)$] are 3x1; [$x_{31}(t)$] and [$e_{31}(t)$] are 4x1.

The matrices [A_3] and [B_3] are sparse, as the matrix [A_3] is almost singular. The following submatrices are also null: [a_{13}]; [a_{23}]; [a_{31}]; [a_{32}]; [b_{23}]; [b_{32}]; [b_{33}]; [$e_{31}(t)$]. In addition, the submatrices [a_{33}], [b_{13}], [b_{31}] are diagonal. The submatrices of the state variables have the following form:

$$\left[x_{11}(t)\right]=\begin{bmatrix} i'_1(t) \\ i'_2(t) \\ \vdots \\ i'_5(t) \end{bmatrix};\ \left[x_{21}(t)\right]=\begin{bmatrix} i'_6(t) \\ i'_8(t) \\ i'_9(t) \end{bmatrix};\ \left[x_{31}(t)\right]=\begin{bmatrix} u_{cs1}(t) \\ u_{cs2}(t) \\ u_{cs3}(t) \\ u_{cs4}(t) \end{bmatrix}. \qquad (14)$$

The algorithm to avoid the fact, that the matrix [A_3] is almost singular, consists in breaking equation (11) into three matrix equations containing the above submatrices, performing at each time step forward a similar computational procedure as in the second interval described by equations (6), (7), (8), (9) and (10).

Fig. 3 shows the plots of the investigated quantities in the three intervals of the switching process of the ADAVR with active-inductive load (*R_L=30,231Ω and L_L=2,0951H*). The phase of commutation of the switch K_4 is chosen to be φ=225°.

Fig. 3a shows the input current $i_1(t)$ as a function of time; Fig. 3b shows the output current $i_2(t)$; Fig. 3c – the current $i_4(t)$ through the switch K_4; Fig. 3d – the voltage drop $u_{cs3+rs3}(t)$ upon the third switch-off C_{S3}+r_{S3}; Fig. 3e – the current $i_{cs4+rs4}(t)$ through the fourth switch-off group and Fig. 3f – the voltage drop $u_{cs4+rs4}(t)$ over the fourth switch-off group C_{S4}+r_{S4}.

In Table 1a, 1b and 1c are presented the experimental data and computer simulations using the proposed algorithm and the program AVTO for a specific ADAVR at an input voltage of 160V with three different active-inductive loads.

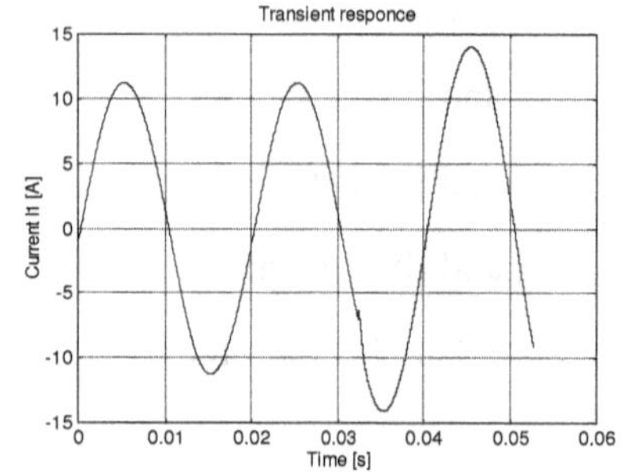

a)

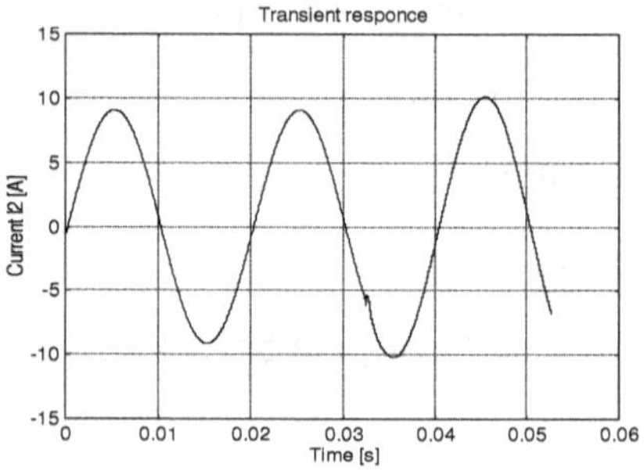

b)

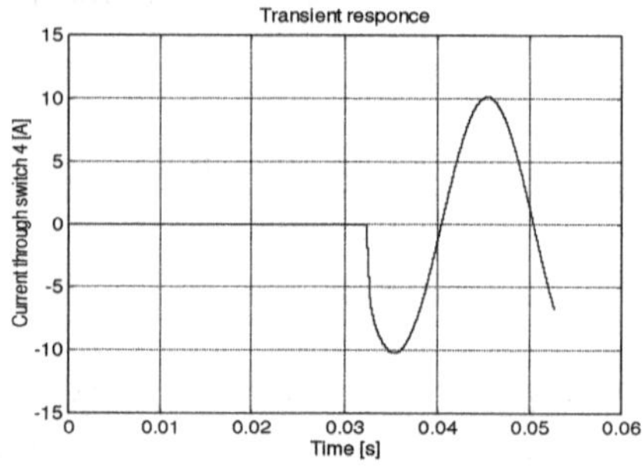

c)

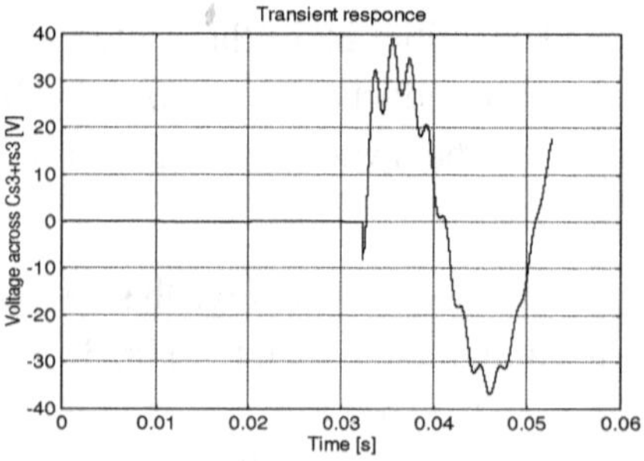

d)

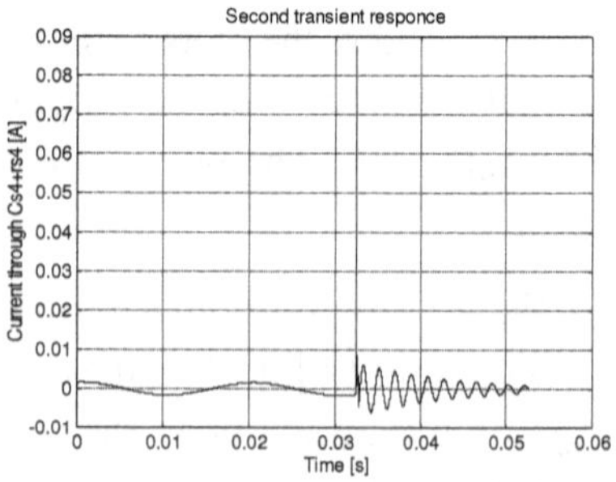

e)

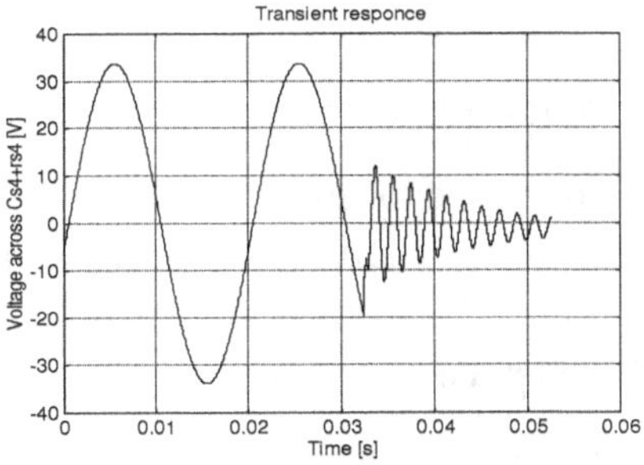

f)

Fig. 3. Graphs of the investigated quantities in the ADAVR with active-inductive load for the three intervals of the switching process.

Table 1. Results of physical experiments and computer simulations.

a) L_L=1,0460H

	Results from the physical experiment					*Results from the computer simulations*				
R_L	**U_1**	**I_1**	**I_0**	**I_2**	**U_2**	**U_1**	**I_1**	**I_0**	**I_2**	**U_2**
Ω	V	A	A	A	V	V	A	A	A	V
115,8	160	3	0,9	2,05	220,3	160	2,8328	0,8501	1,9918	218,63
61,8	160	5,25	1,43	3,55	220	160	5,0637	1,4422	3,6275	218,40
35,5	160	9,75	2,7	7,03	219,9	160	8,6085	2,3949	6,2175	217,94

b) L_L=2,4933H

	Results from the physical experiment					*Results from the computer simulations*				
R_L	**U_1**	**I_1**	**I_0**	**I_2**	**U_2**	**U_1**	**I_1**	**I_0**	**I_2**	**U_2**
Ω	V	A	A	A	V	V	A	A	A	V
115,8	160	3,15	0,98	2,2	220,3	160	3,0062	0,9106	2,1020	217,69
61,8	160	5,1	1,49	3,71	220,1	160	5,1673	1,4798	3,6921	217,46
35,5	160	10,05	2,75	7,2	220	160	8,6598	2,4150	6,2482	217,00

c) L_L=11,356H

	Results from the physical experiment					*Results from the computer simulations*				
R_L	**U_1**	**I_1**	**I_0**	**I_2**	**U_2**	**U_1**	**I_1**	**I_0**	**I_2**	**U_2**
Ω	V	A	A	A	V	V	A	A	A	V
115,8	160	2,72	0,83	1,95	220	160	2,7319	0,8131	1,9301	219,16
61,8	160	4,8	1,39	3,5	220,1	160	4,9879	1,4156	3,5793	218,94
35,5	160	9,6	2,65	7	219,9	160	8,553	2,3761	6,1813	218,48

III. CONCLUSIONS

In conclusion, the following conclusions are drawn:

1. A detailed algorithm is proposed for solving the system of equations describing the electrical equilibrium in the ADAVR under active-inductive load, which allows overcoming the peculiarities of the systems matrices in the second and third interval of the voltage regulator action.

2. On the basis of the proposed algorithm, the AVTO computer program is extended, covering the specifics of the circuit solution, the autotransformer parameters, the size and type of the load, the angle of commutation and the parameters of the thyristor switches used. The shown graphs allow the evaluation of the nature and duration of the commutation process, as well as the electrical load of the individual circuits and the semiconductor commutators.

3. In the tables shown, the comparison between the data from computer simulations and physical experiments with the prototype of the ADAVR with the investigated loads is presented, which indicates a good coincidence of the obtained results.

References

[1]. Barudov S., Barudov E., Discrete alternating current regulators and stabilizers, PENSOFT Sofia-Moscow 2006.

[2]. Lipkovskiy K. Executive structures of AC voltage converters based on transformer switches, Naukova Dumka, 1983.

[3]. Barudov S., Panov E., Study of the loading of the switching elements in a step voltage regulator, Annual Proceedings of Technical University of Varna, Varna, 2004, pp. 117÷122.

[4]. Barudov E., Barudov S., Panov E., Switching processes in a step voltage regulator, Acta Universitatis Pontica Euxinus, Volume 4, Number 1, 2005, pp. 21÷25.

[5]. Barudov S., Panov E., Barudov E., Parameters of the switching process in a step-up AC voltage regulator, SPB: PEPIK, Methods and means of estimation of power equipment condition, Vol. 25, 2005, pp. 67÷76.

[6]. Panov E., Barudov E., Barudov S., Advanced algorithm for the analysis of a precise non-linear model of an autotransformer discrete voltage regulator with semiconductor commutation elements, Annual of TU-Varna, 2009, Bulgaria, pp. 47÷52, ISSN: 1311-896X.

[7]. Barudov E., Panov E., Barudov S., Analysis of electrical processes in alternating voltage control systems, Journal of international scientific publication, Volume IV, Part 1, 2010, pp. 154÷182, ISSN: 1313 2539.

Contact:

1. Emil Barudov, Technical University - Varna, 9010, Bulgaria, 1 Studentska str., Faculty of Electrical Engineering, Department TIE, e-mail: ugl@gyuvetch.bg

2. Emil Panov, Technical University - Varna, 9010, Bulgaria, 1 Studentska str., Faculty of Electrical Engineering, Department TIE, e-mail: eipanov@yahoo.com

Reviewer:

Prof. Ph.D. Eng. Mat. K. Gerasimov

Chapter 3. Comparison Between the Approximate and Precise Models of the ADAVR

Vector Analysis and Comparative Valuation of Precise and Approximate Non-Linear Models of Discrete Regulator with Reducing Input AC Voltage

Emil Panov[1], Emil Barudov[2] and Stefan Barudov[3]

(Published in the Proceedings of XLVIII-th International Scientific Conference on Information, Communication and Energy Systems and Technologies ICEST'2013, 26-29 June 2013, Ohrid, Macedonia, Volume 2, pp.771-774, ISBN 978-9989-786-89-1.)

[1]Emil Panov (Assoc. Prof., PhD, Eng.) is with the Electrical Faculty at the Technical University of Varna, 1 Studentska str., Varna 9010, Bulgaria, e-mail: eipanov@yahoo.com.
[2]Emil Barudov (Assist. Prof., Eng.) is with the Electrical Faculty at the Technical University of Varna, 1 Studentska str., Varna 9010, Bulgaria, e-mail: ugl@abv.bg.
[3]Stefan Barudov (Prof., DSc, Eng.) is with the Engineering Faculty at the Naval Academy "N. Vaptsarov" Varna, 73 Vasil Drumev str., Varna 9026, Bulgaria e-mail: sbarudov@abv.bg.

Abstract – **The discrete regulation of AC voltages is used in many cases. Most often, it is achieved by power electronic converters, based on a transformer (autotransformer) and switching by the means of controllable semiconductor switches.**

The present work is dedicated to the vector analysis and to the vector measurements in an autotransformer discrete voltage regulator, replaced by precise and approximate non-linear models with changing input voltage and commutation angle of the semiconductor switches.

Keywords – **vector analysis, vector measurements, voltage regulator, semiconductor switch, thyristor**

I. Introduction

In Fig. 1 the equivalent circuit of an autotransformer discrete alternating voltage regulator (ADAVR) with four terminals and four controllable semiconductor switches (CSS) is shown. The equivalent circuit corresponds to the approximate model as the losses in the autotransformer core are not considered, the semiconductor switches are accepted for ideal and the RC groups, which shunt the thyristors in the semiconductor switches, are not taken into account. The adopted control algorithm is connected with switching at random moment as the commutation is always performed between two neighbouring semiconductor switches [1, 2].

The feeding with control pulses is suspended to the thyristor switch, which will be turned off (the switch remains conductive until the natural commutation of the thyristors inside it), and the other switch starts to be fed with control pulses.

The equivalent circuit of the precise model and its mathematical description are presented in details in literature [3, 4].

The aim of the current work is the comparative research into the loading (the semiconductor switches, the autotransformer windings) by vector analysis and vector measurements of the quantities during the commutation process at different character of the load – R, RL and RC and different commutation angles.

II. Analysis

In this paper a vector analysis at low values of the input voltage of ADAVR is conducted. In this case, the switch K_4 is turned off and the switch K_3 is switched on (Fig. 1).

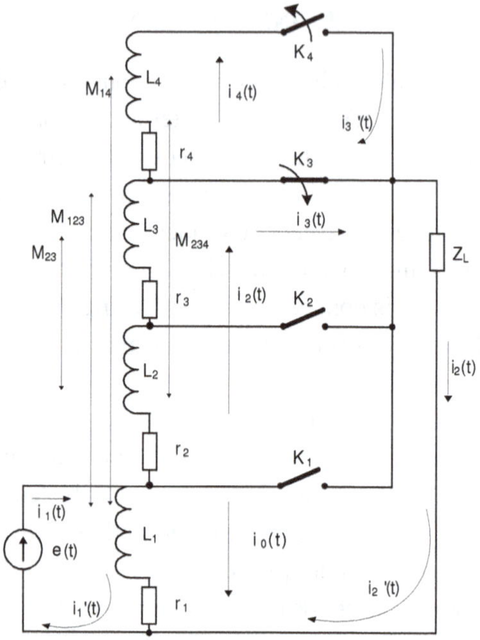

Fig. 1. Equivalent circuit of an ADAVR with four CSS at input voltage change.

At the same time, at certain angles φ of the commutation process, it is possible r_4 and L_4 (Fig. 1) to be connected in short circuit through two thyristors, one from the switch K_4 and one from the switch K_3 respectively, for the time until the natural commutation of the switch K_4 occurs. This mode assumes an overload regime of the fourth section of the autotransformer and the two connected in series thyristors from the switches K_4 and K_3, when they are triggered.

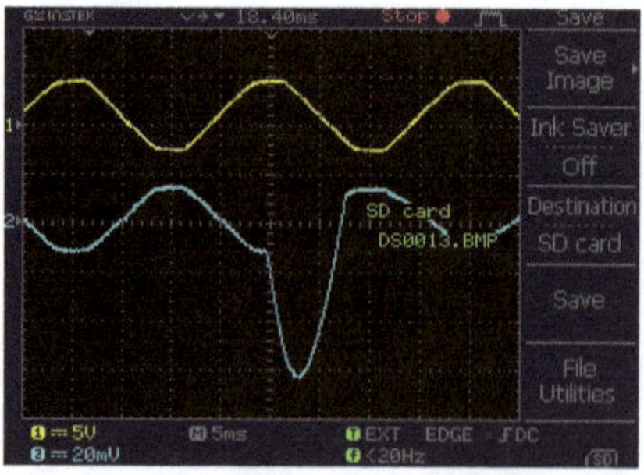

a) Oscillogram of the input current $i_1(t)$ at $\varphi=270^0$.

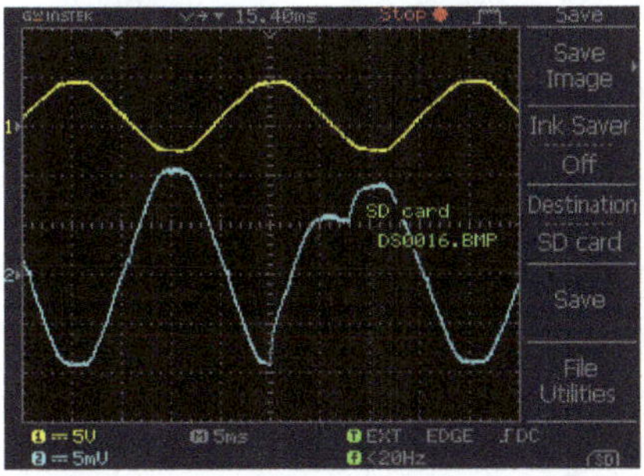

b) Oscillogram of the output current $i_2(t)$ at $\varphi=270^0$.

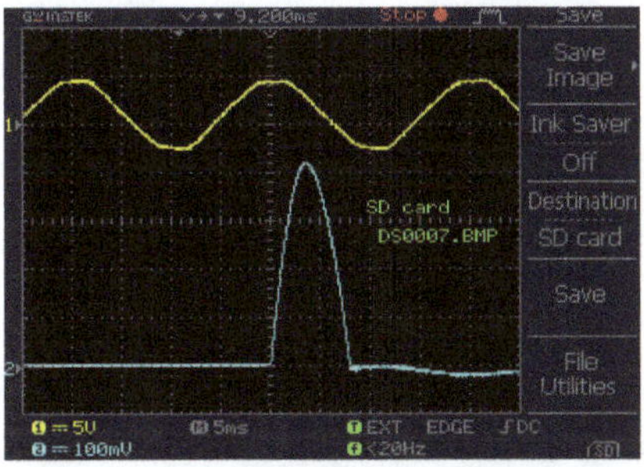

c) Oscillogram of the current through switch K_3 - $i_3(t)$ at $\varphi=270^0$.

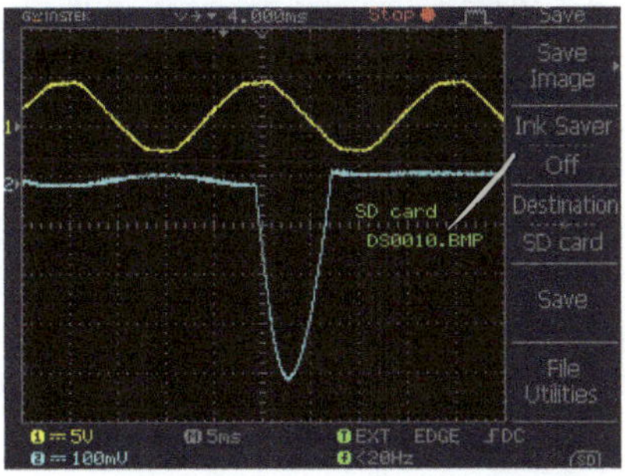

d) Oscillogram of the current through switch K_4 - $i_4(t)$ at $\varphi=270^0$.

Fig. 2. Experimental oscillograms of the currents of the ADAVR for R load and commutation angle $\varphi=270^0$.

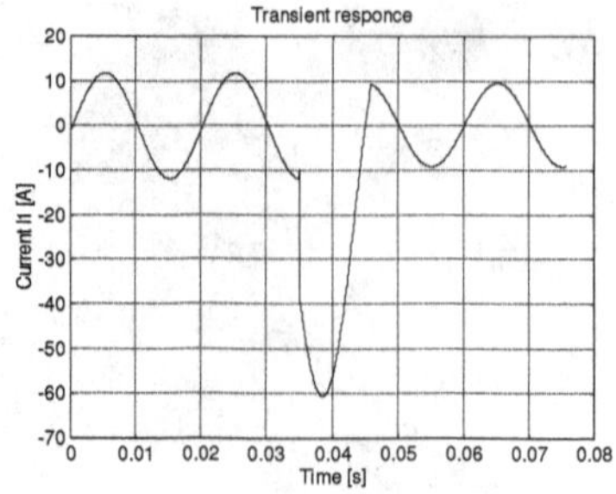

a) Simulation of the input current $i_1(t)$ at $\varphi=270^0$.

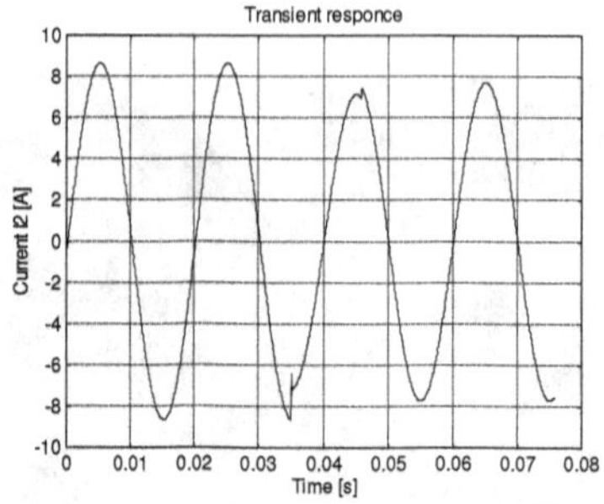

b) Simulation of the output current $i_2(t)$ at $\varphi=270^0$.

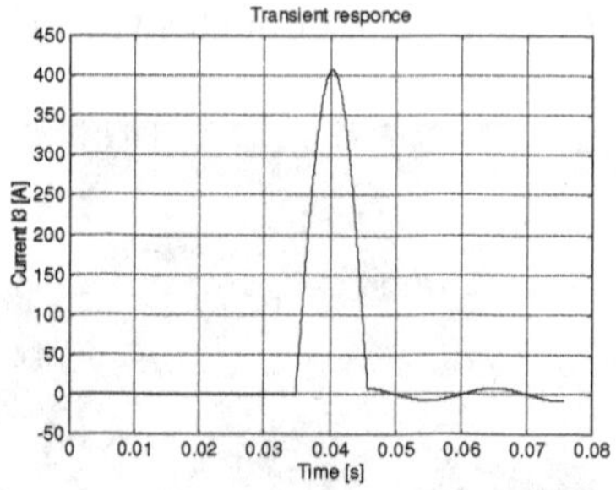

c) Simulation of the current through switch K_3 - $i_3(t)$ at $\varphi=270^0$.

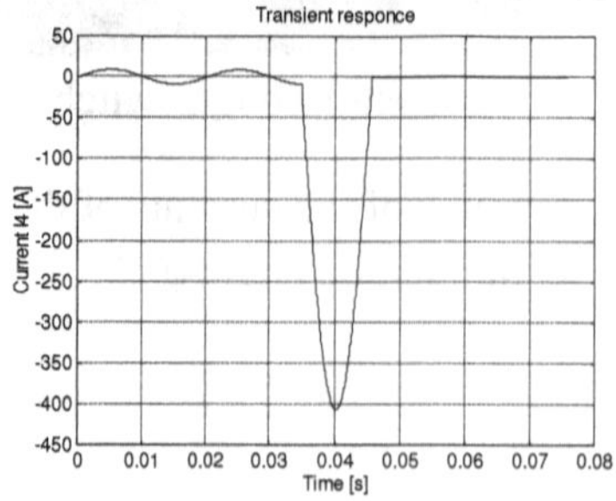

d) Simulation of the current through switch K_4 - $i_4(t)$ at $\varphi=270^0$.

Fig.3. Computer simulation of the currents of the ADAVR for R load and commutation angle $\varphi=270^0$.

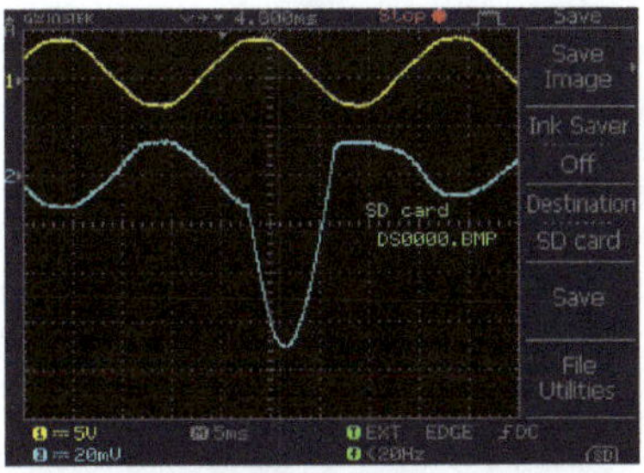

a) Oscillogram of the input current $i_1(t)$ at $\varphi=225^0$

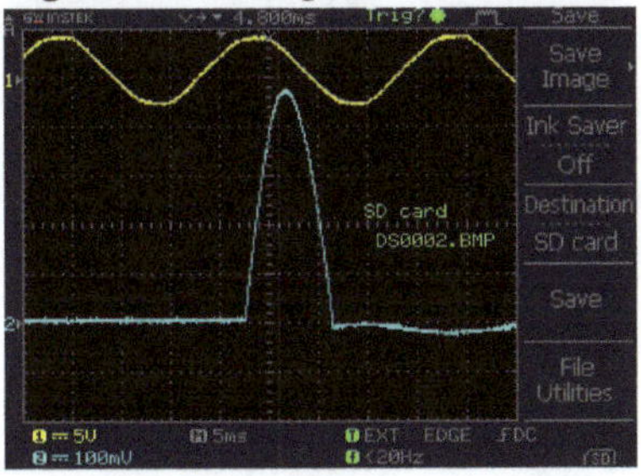

b) Oscillogram of the current through switch K_3 - $i_3(t)$ at $\varphi=225^0$.

Fig. 4. Experimental oscillograms of the currents of the ADAVR for RL load and commutation angle $\varphi=225^0$.

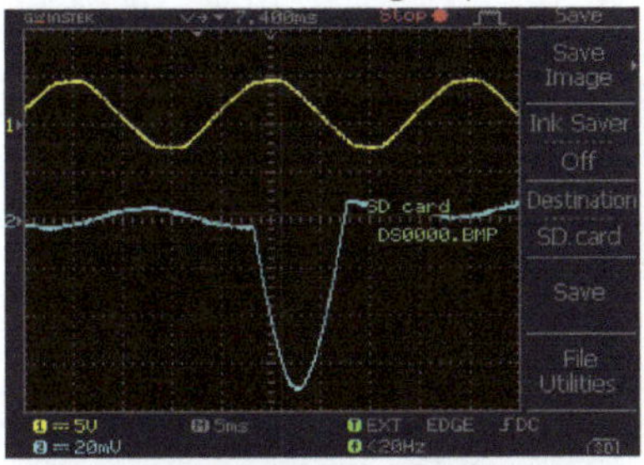

a) Oscillogram of the input current $i_1(t)$ at $\varphi=225^0$.

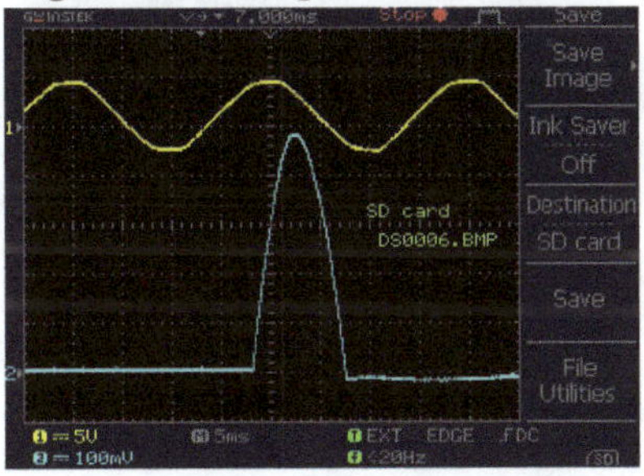

b) Oscillogram of the current through switch K_3 - $i_3(t)$ at $\varphi=225^0$.

Fig. 5. Experimental oscillograms of the currents of the ADAVR for RC load and commutation angle $\varphi=225^0$.

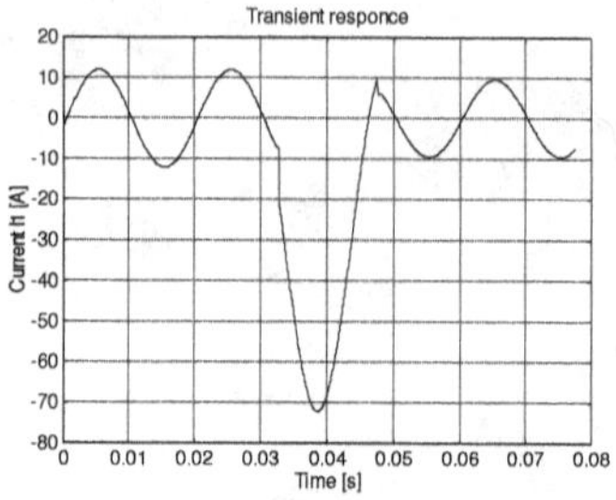

a) Simulation of the input current $i_1(t)$ at $\varphi=225^0$.

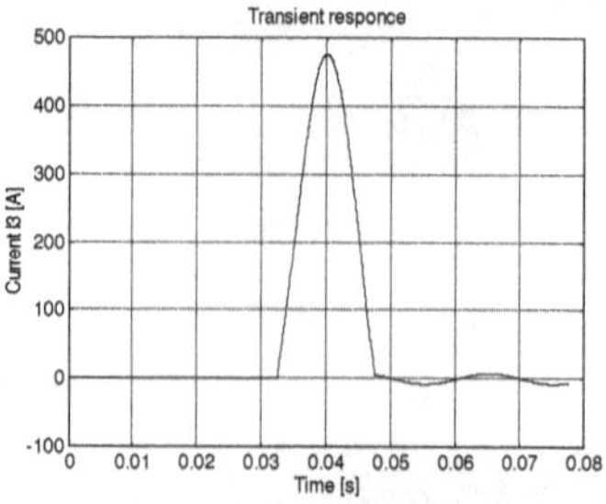

b) Simulation of the current through switch K_3 - $i_3(t)$ at $\varphi=225^0$.

Fig. 6. Computer simulation of the currents of the ADAVR for RL load and commutation angle $\varphi=225^0$.

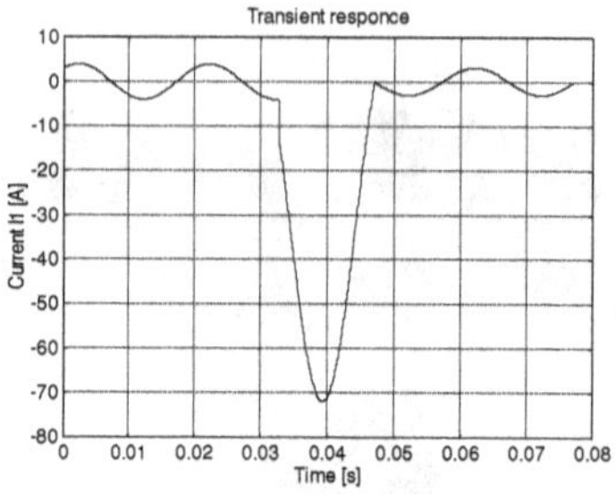

a) Simulation of the input current $i_1(t)$ at $\varphi=225^0$.

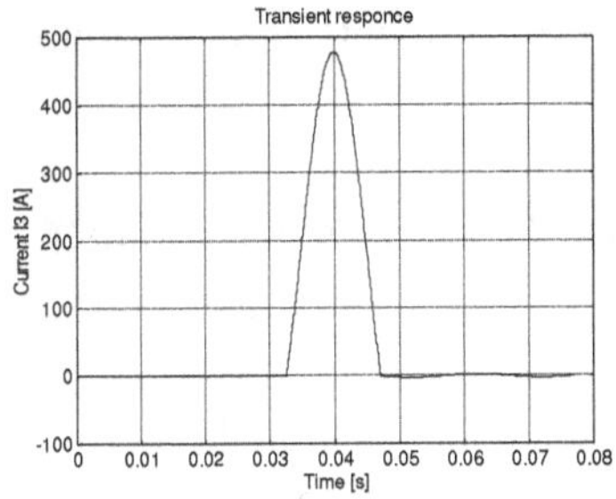

b) Simulation of the current through switch K_3 - $i_3(t)$ at $\varphi=225^0$

Fig. 7. Computer simulation of the currents of the ADAVR for RC load and commutation angle $\varphi=225^0$.

In case of an active load (R-load) and commutation angle $\varphi=270^0$, the results from the experimental oscillograms and the computer simulations with the programme AVTO in MATLAB integrated environment, are presented in Fig. 2 and Fig. 3. When we have RL and RC loads, the development of the commutation process can be followed in Fig. 4 and Fig. 5, Fig. 6 and Fig. 7, respectively. The experimental oscillograms and simulations are taken at commutation angle $\varphi=225^0$.

The received experimental data for a concrete regime (closed switch K_4 and opened switch K_3) are shown in Table 1. The analytical data in the table are very close to those from the analysis of the precise model [4, 5, 6]. The numerical values from the simulations with the approximate model of the voltage regulator differ with average deviation 0,9% (minimum deviation 0,07% and maximum deviation 2,83%) from those, received by a simulation with the precise model.

Table 1

Load	*Experimental data*			*Computer simulations*		
	I_1	I_2	U_2	I_1	I_2	U_2
	A	A	V	A	A	V
R	8,45	6,16	219	8,3965	6,1205	218,6
RL	9,75	7,03	219,9	8,5181	6,2016	217,8
RC	2,85	2,15	220,3	2,8034	2,1421	222,6

• *Note:* The loads for the experiments and the simulations, shown in the table, are as follows:
R load - 35,7Ω;
RL load - R-35,7Ω, L-1,76H (connected in parallel);
RC load - R-61,78Ω, C-38,09μF (connected in series).

In Table 2 the vector quantities of the currents and the voltages of the autotransformer discrete regulator from the equivalent circuit in Fig. 1 are presented.

Table 2

Load	***Results from vector measurements***		
	$\dot{I}_1$	$\dot{I}_2$	$\dot{U}_2$
	A	A	V
R	$8{,}45e^{-j0{,}08}$	$6{,}16e^{-j0{,}06}$	219
RL	$9{,}75e^{-j0{,}14}$	$7{,}03e^{-j0{,}12}$	$219{,}9e^{j0{,}06}$
RC	$2{,}85e^{j0{,}89}$	$2{,}15e^{j0{,}92}$	$219e^{-j0{,}01}$

III. CONCLUSION

A vector analysis and vector measurements of the quantities, referred to the commutation and the regimes in a power semiconductor converter with discrete regulation of the input AC voltage magnitude to the joined consumers have been conducted. A precise and approximate (with certain simplifications) models have been examined at different angles for the commutation process and for different loads.

Both results – from the experiments and from the computer simulations, show a very good match of the obtained results.

A program AVTO in MATLAB is developed, and it allows visualization of the computer simulations as well as examination of the discrete AC voltage regulator with different loads and parameters of the commutation processes.

ACKNOWLEDGEMENT

The presented results in the current paper are obtained under working at project NP1/2013 of the Technical University of Varna, Bulgaria, funded by the National Budget of Republic of Bulgaria.

References

[1] Harlow James H., Transformers. The Electric Power Engineering Handbook. Ed. L.L. Grigsby Boca Raton: CRC Press LLC, 2001.

[2] Fernando S., Power Electronics Handbook – voltage regulators. (Third Edition), 2011, Sónia Ferreira Pinto.

[3] Barudov E., Panov E., Barudov S., Analysis of Electrical Processes in Alternating Voltage Control Systems. Journal of International Scientific Publication: Materials, Methods & Technologies, 2010, Vol. 4, Part 1, pp. 154÷182, ISSN 1313 2539.

[4] Barudov E., Panov E., Barudov S., Exploration of Precise Non-Linear Model of Discrete Autotransformer Step-Voltage AC Regulator with Semiconductor Commutators. Annual of TU-Varna, 2007, Bulgaria, pp. 3÷10, ISSN: 1311-896X.

[5] Barudov E., Panov E., Barudov S., Analysis of Electrical Processes in a Discrete Alternating Voltage Regulator with Active-Capacitive Load. International Scientific and Technical Conference "Electrical Power Engineering 2010", pp. 332÷341, ISBN 978-954-20-0497-4.

[6] Barudov E., Panov E., Barudov S., Analysis of Electrical Processes in a Discrete Alternating Voltage Regulator with Active-Inductive Load. Annual of TU-Varna, 2010, Bulgaria, pp. 30÷35, ISSN: 1311-896X.

Chapter 4. Exploration on the Commutation Regimes of the ADAVR

COMMUTATION PROCESSES WITH REACTIVE LOADS IN A DISCRETE ALTERNATING VOLTAGE REGULATOR

Emil Barudov, Emil Panov, Stefan Barudov

(Published in the Annual of the Technical University of Varna, Vol. I, 2011, pp. 3 - 8, ISSN: 1311-896X. (in Bulgarian))

Abstract: In networks with limited power, the voltage supply of the separate consumers can be controlled in admissible range through autotransformer discrete alternating voltage regulators (ADAVR). This paper examines the switching processes occurring at the commutation of the thyristor switches of ADAVR Different active, active-capacitive and active-inductive loads at different angles of commutation are studied. The performed experimental verification with the prototype reveals good correspondence with the data from the computer simulations.

Keywords: autotransformer, discrete alternating voltage regulator, semiconductor commuting elements, thyristor, commutation process

I. INTRODUCTION

There are electrical networks in which the phase voltage varies beyond permissible limits. The reasons for this are complex, but in order to ensure trouble-free operation of the consumers, the magnitude of the input supply voltage must be maintained within the limits specified by Bulgarian State Standard 50160. One method of doing this is through the use of autotransformer discrete alternating voltage regulators (ADAVR) with semiconductor commutation elements (SCE) [1, 2, 3]. The design of such regulators requires the use of tools for computer analysis and simulation of the complex processes occurring in these devices [3, 4]. The main problem for the creation of accurate models and analysis algorithms, which allow accurate simulations of the responses of ADAVR, under different operating modes appear to be the complexity of the switching processes and the change of the nature and magnitude of the load. The work compares analytically and experimentally obtained results for different loads and angles of commutation.

II. ANALYSIS

The regulation of the amplitude variation range of the supply voltage of the consumers is realized by the automatic switching of the transformer terminals (auto-transformer).

Fig. 1 shows the scheme of the ADAVR with four SCE. The scheme includes the parameters of the switching groups, the switch-off groups, the magnetic circuit, the parameters of the individual winding sections and the existing non-linearities.

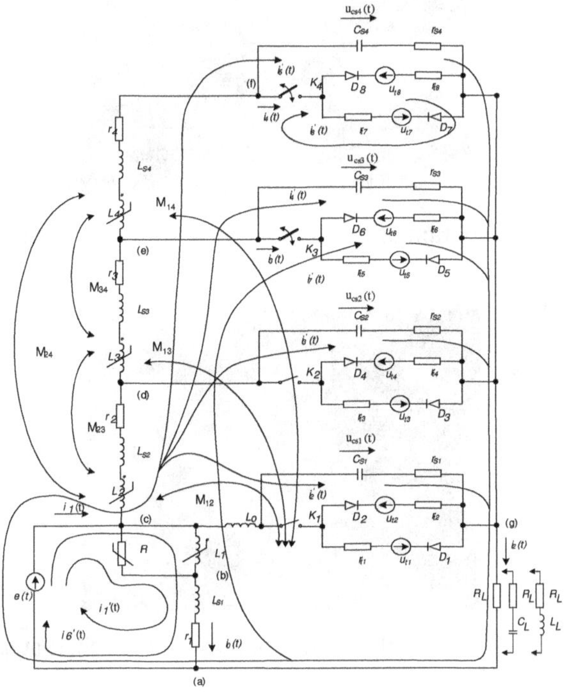

Fig. 1. Replacement scheme of ADAVR with four SCE.

During the researches are observed the most severe mode of operation - commutation from the penultimate to the last boost stage of the ADAVR - the input voltage decreases from 170V to 160V. The commutation process covers three intervals [1, 2]. In the first interval, switch K3 is closed and all others are open; in the second interval, switches K3 and K4 are closed; and in the third interval, only switch K4 is closed.

Fig. 2a, 2b, and 2c show the control pulses applied to the photo-controlled symistor in the SCE [5] at commutation phases $\varphi = 225^0$, 270^0 and 315^0.

In Fig. 3a, 3b, 4a, 4b, 5a, 5b, 6a and 6b the input current of the voltage regulator, $i_1(t)$, the current through SCE - K3 – $i_{s3}(t)$, the current through SCE - K4 - $i_{ks4}(t)$ and the current through the load $i_2(t)$, respectively, are shown. Index "a" shows the results obtained experimentally, and index "b" shows the analytical results obtained from the mathematical modeling process with the developed MATLAB program AVTO.

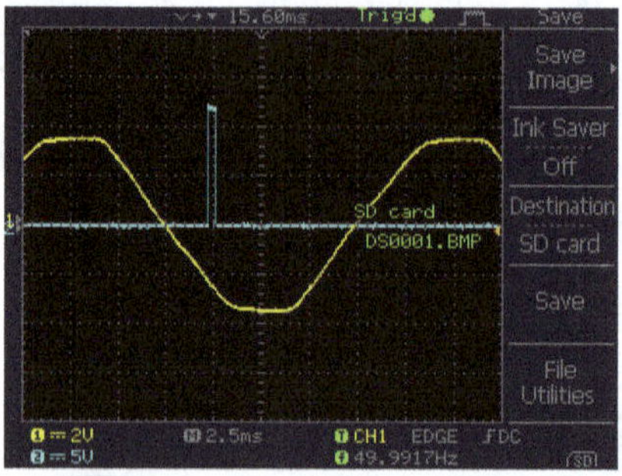

Fig. 2a. Oscillogram of the control pulses of the SCE at φ=225^0.

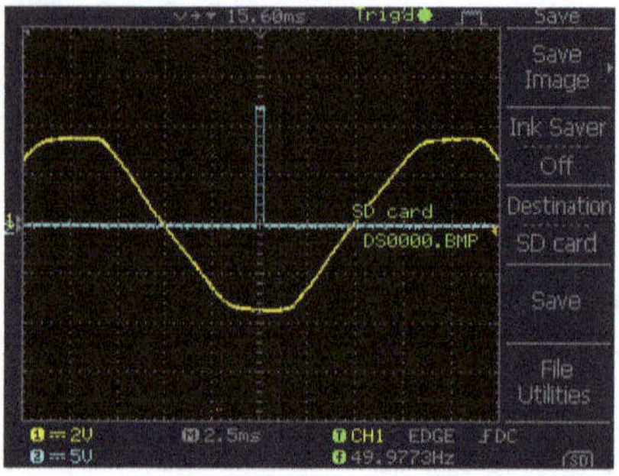

Fig. 2b. Oscillogram of the control pulses of the SCE at φ=270^0.

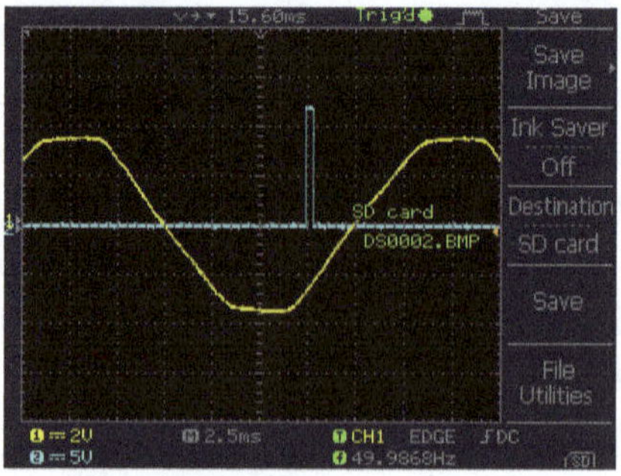

Fig. 2c. Oscillogram of the control pulses of the SCE at φ=315^0.

The results shown on the oscillograms were obtained with input supply voltage of the ADAVR E=160V; R_L=35,5Ω and initial phase of the commutation φ=270^0.

In Fig. 7a, 7b; 8a, 8b; 9a, 9b, 10a and 10b the input current-$i_1(t)$, the current through the SCE - K3- $i_{s3}(t)$ the current through the SCE - K4 - $i_{s4}(t)$ and the current through the load $i_2(t)$ are shown respectively at the input supply voltage of the ADAVR E=160V, L_L=2,1712H; and R_L=60,6Ω, connected in parallel. The initial phase of commutation is φ=225^0. Fig. 11a, 11b, 12a, 12b, 13a, 13b, 14a and 14b show respectively the input current of the voltage regulator $i_1(t)$, the current through the SCE - K3 - $i_{s3}(t)$, the current through the SCE - K4 - $i_{s4}(t)$ and the current through the load $i_2(t)$, at input supply voltage of the ADAVR E=160V, C_L=38,09μF and R_L=61,8Ω, connected in series. The initial commutation phase is φ=225^0.

On all oscillograms showing the currents in the regulator, 1mV/div corresponds to 1A/div.

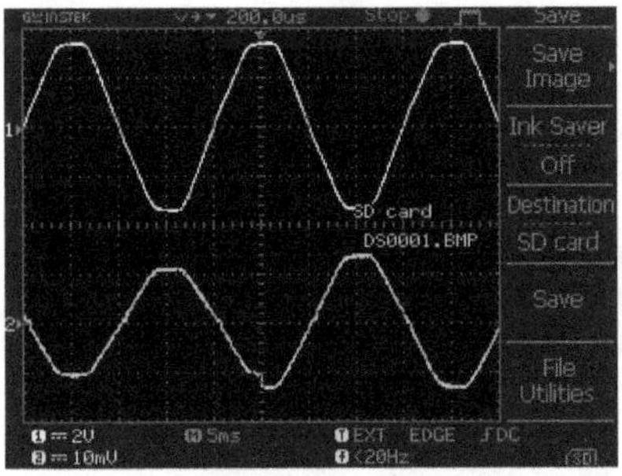

Fig. 3a. Oscillogram of $i_1(t)$.

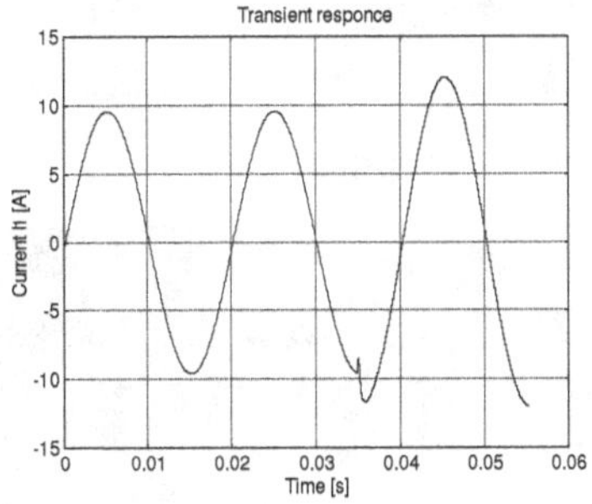

Fig. 3b. Comp. simulation of $i_1(t)$.

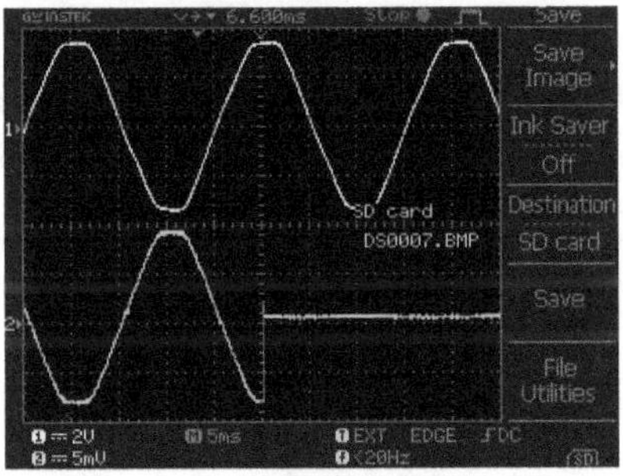

Fig. 4a. Oscillogram of $i_{s3}(t)$.

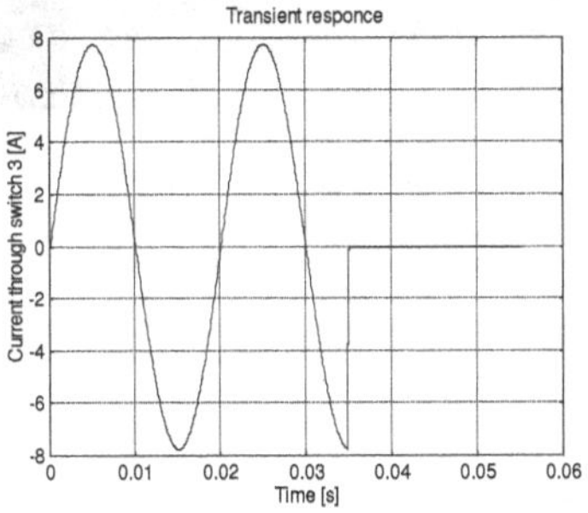

Fig. 4b. Comp. simulation of $i_{s3}(t)$.

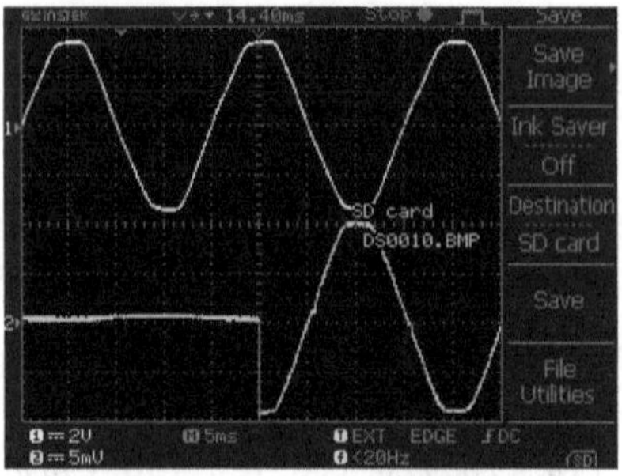

Fig. 5a. Oscillogram of $i_{s4}(t)$.

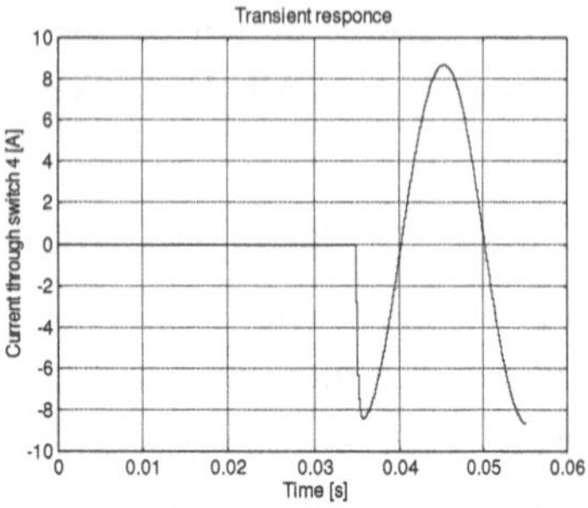

Fig. 5b. Comp. simulation of $i_{s4}(t)$.

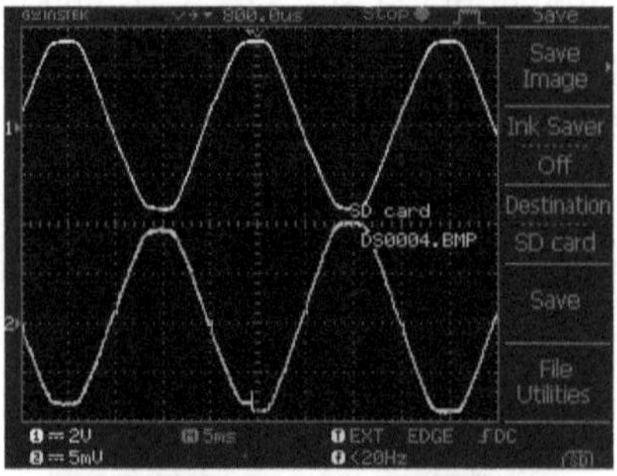

Fig. 6a. Oscillogram of $i_2(t)$.

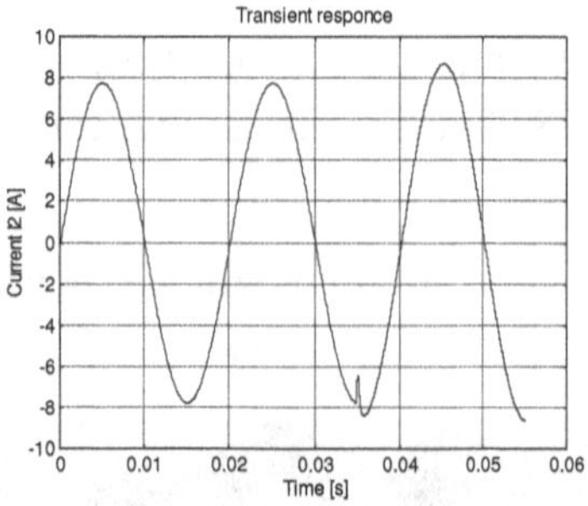

Fig. 6b. Comp. simulation of $i_2(t)$.

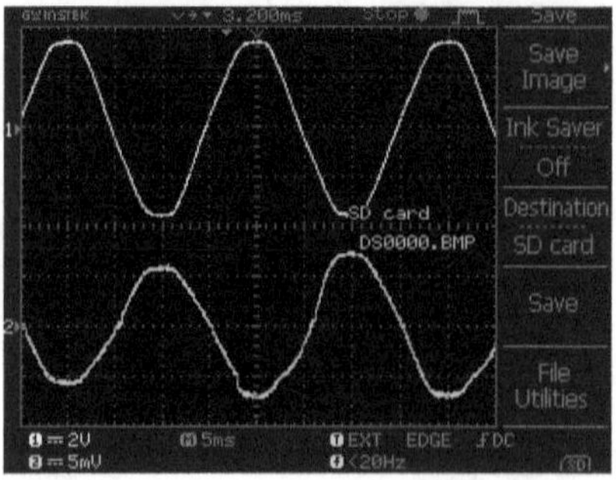

Fig. 7a. Oscillogram of $i_1(t)$.

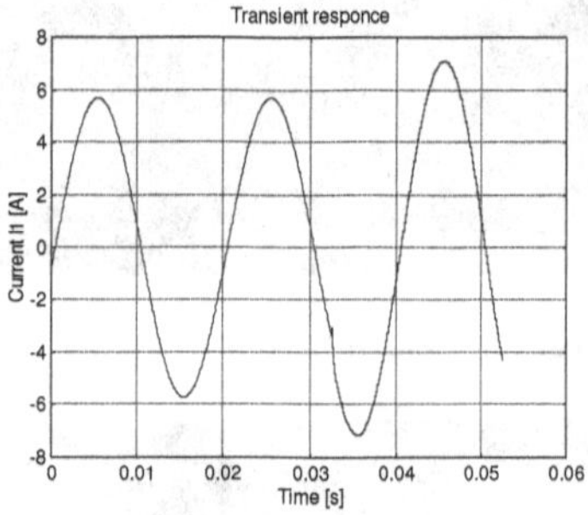

Fig. 7b. Comp. simulation of $i_1(t)$.

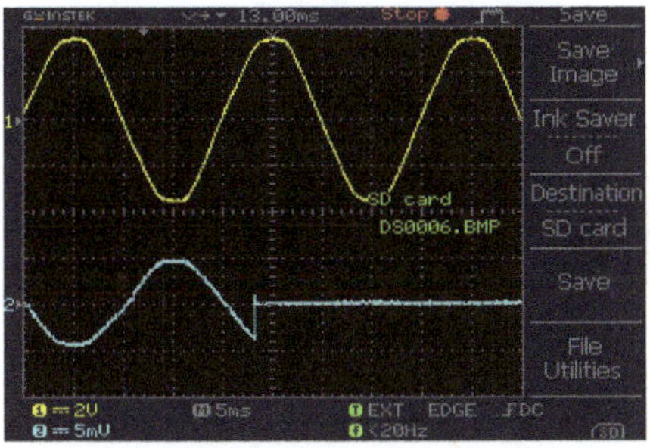

Fig. 8a. Oscillogram of $i_{s3}(t)$.

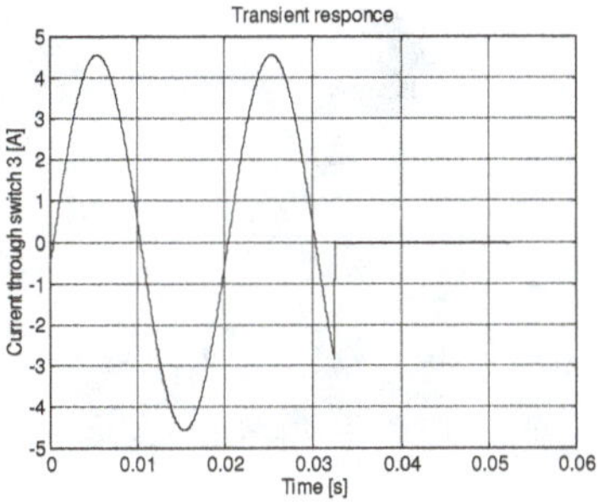

Fig. 8b. Comp. simulation of $i_{s3}(t)$.

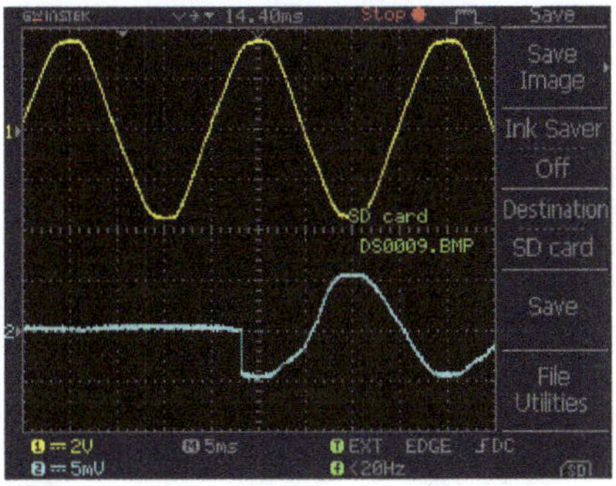

Fig. 9a. Oscillogram of $i_{s4}(t)$.

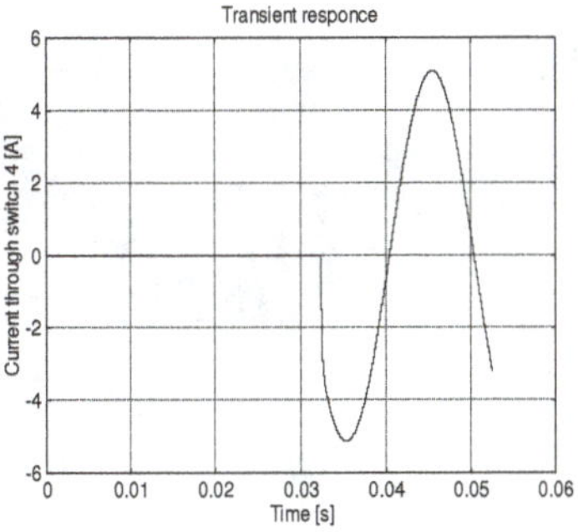

Fig. 9b. Comp. simulation of $i_{s4}(t)$.

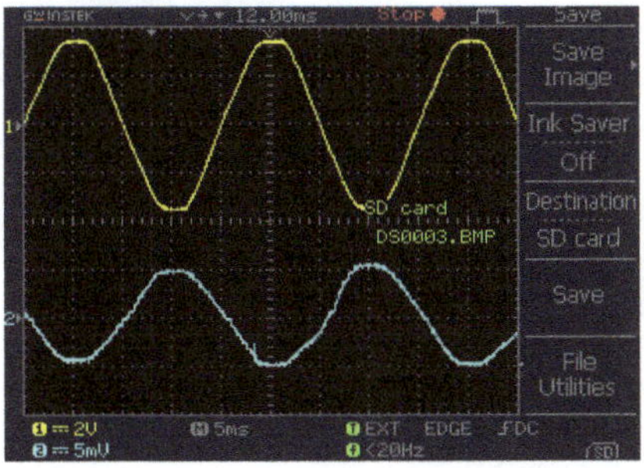

Fig. 10a. Oscillogram of $i_2(t)$.

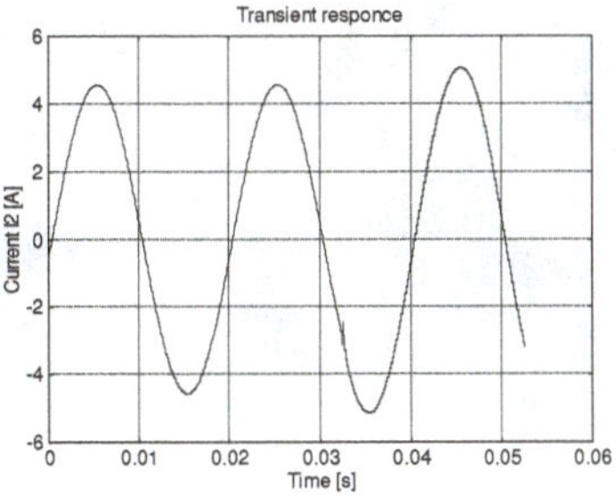

Fig. 10b. Comp. simulation of $i_2(t)$.

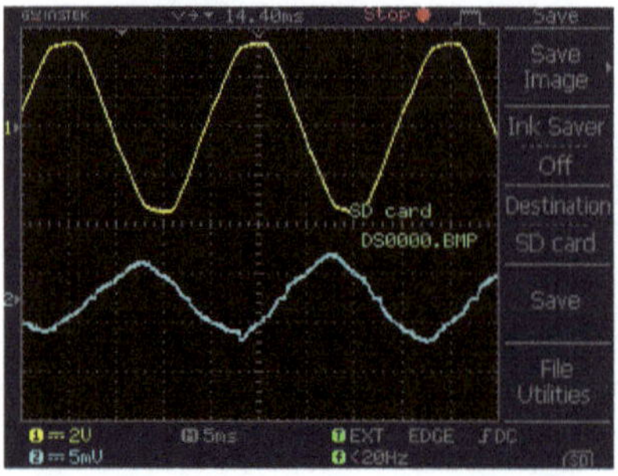

Fig. 11a. Oscillogram of $i_1(t)$.

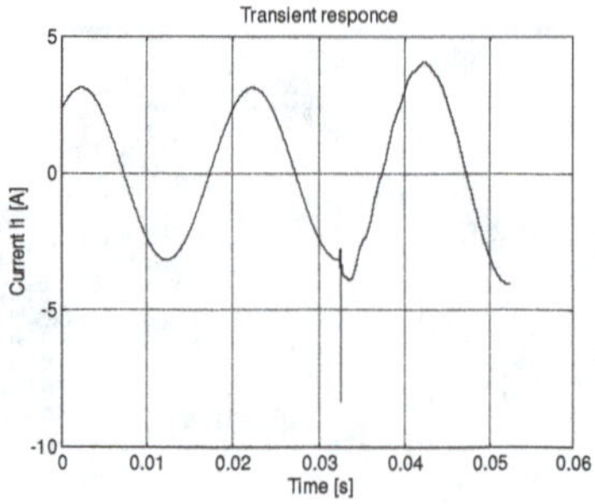

Fig. 11b. Comp. simulation of $i_1(t)$.

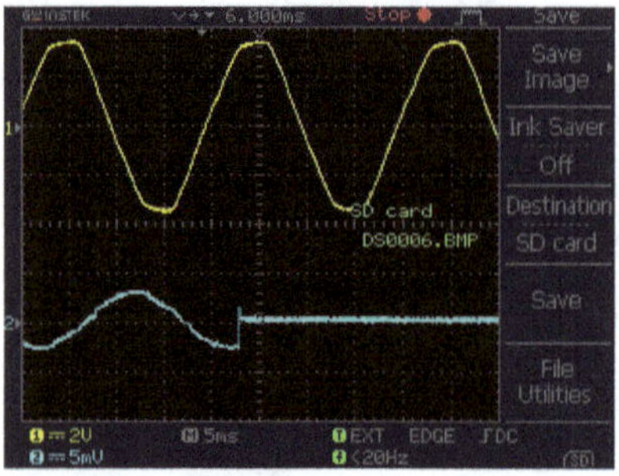

Fig. 12a. Oscillogram of $i_{s3}(t)$.

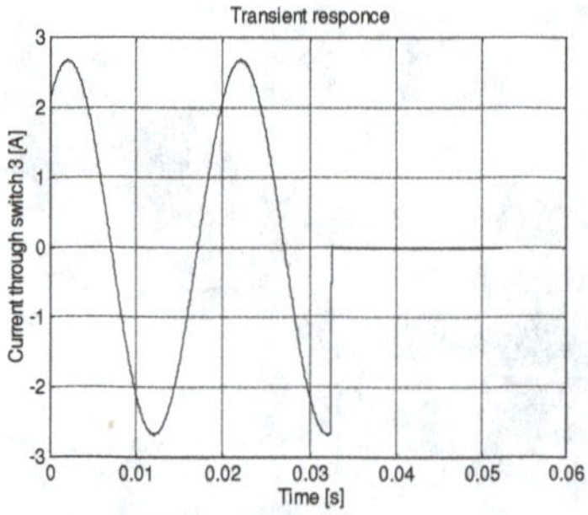

Fig. 12b. Comp. simulation of $i_{s3}(t)$.

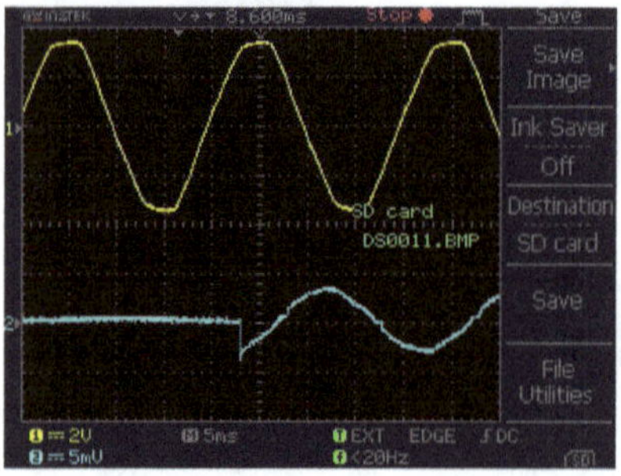

Fig. 13a. Oscillogram of $i_{КЛ4}(t)$.

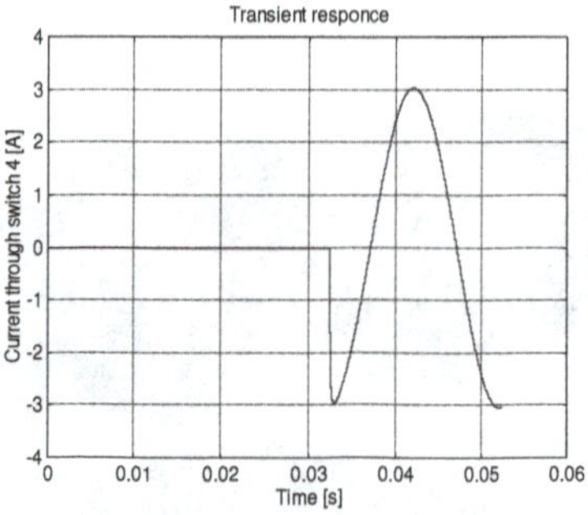

Fig. 13b. Comp. simulation of $i_{s4}(t)$.

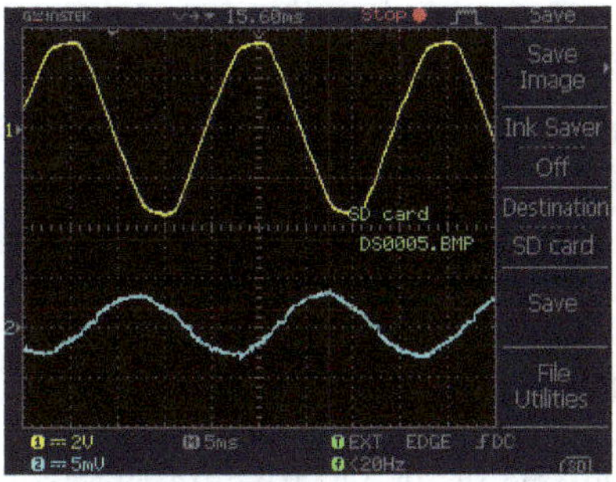

Fig. 14a. Oscillogram of $i_2(t)$.

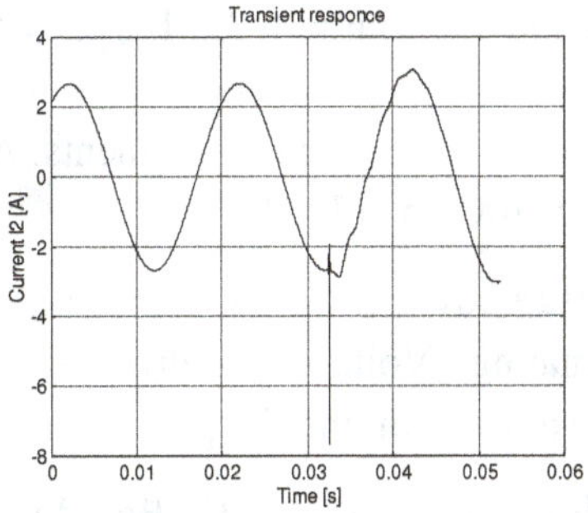

Fig. 14b. Comp. simulation of $i_2(t)$.

III. CONCLUSIONS

1. Computer simulations and experimental investigations have been carried out to develop an accurate model of the ADAVR, taking into account the specificity of the implementation of the SCE, the nonlinearities in the magnetic circuits and covering the parameters such angle of commutation, nature and magnitude of the load:

- More accurate quantitative data on the duration of the switching process and the load on the switches during its implementation were obtained;

- The influence of the switching process on the input current i.e. on the supply network and other consumers is visualized.

2. Experimental investigations have been carried out at the above-mentioned load character and switching phases, showing good agreement of the obtained results, which is an indication of the adequacy of the mathematical model and the computer simulation program AVTO.

3. The presence of complex electrical processes in the switching of the thyristor switches in the ADAVR and their dependence on the nature of the load gives rise to the need for further adaptation of the control circuits to implement the described processes.

REFERENCES:

[1]. Barudov E., Panov E., Barudov S., Study of a precise non-linear model of an autotransformer discrete voltage regulator with semiconductor switching elements, Annual of TU-Varna, 2007, pp. 91÷96.

[2]. Panov E., Barudov E., Barudov S., An improved algorithm for analysis of a precise non-linear model of an autotransformer discrete voltage regulator with semiconductor switching elements, Annual of TU-Varna, 2009, Bulgaria, ISSN: 1311-896X, pp. 47÷53.

[3]. Barudov E., Panov E., Barudov S., Analysis of Electrical Processes in Alternating Voltage Control Systems, Journal of International Scientific Publication, Volume IV, part 1, pp. 154÷182, ISSN: 1313 2539, 2010.

[4]. Barudov E., Panov E., Barudov S. Analysis of the electrical processes in a discrete AC voltage regulator with resistive-capacitive load, Proceedings of the International Scientific and Technical Conference "Electroenergetics 2010", 14-16 October 2010, pp. 332-341.

[5]. Barudov E., Panov E., Barudov S., Comparative research into the commutation processes in a discrete alternating step-voltage regulator, Proceeding of the 13-th International Conference ELMA 2011, 21-22 October 2011, pp. 55-60, ISSN 1313-4965.

CONTACTS:

1. Emil Barudov, Technical University - Varna, 9010, Bulgaria, 1 Studentska str., EF, Department TEEI, e-mail: ugl@gyuvetch.bg

2. Emil Panov, Technical University - Varna, 9010, Bulgaria, 1 Studentska str., EF, Department TEEI, e-mail: eipanov@yahoo.com

3. Stefan Barudov, Technical University - Varna, 9010, Bulgaria, 1 Studentska str., EF, Department PE, e-mail: sbarudov@abv.bg

Reviewer:

Prof. Ph.D. Eng. Mat. K. Gerasimov

STUDY OF THE PECULIARITIES OF THE COMMUTATION PROCESSES IN A DISCRETE ALTERNATING VOLTAGE REGULATOR AT ASYNCHRONOUS SWITCHING MODE

Stefan T. Barudov*, Emil S. Barudov**, Emil I. Panov**

* Naval Academy "N. Y. Vaptsarov", Varna

** Technical University – Varna

(Published on the Marine Scientific Forum, Electronics, Electrical Engineering and Automatics. Informatics, N. Y. Vaptsarov", Vol. 4, 2013, pp. 104-109, ISSN 1310-9278. (in Bulgarian))

Abstract: *In discrete alternating voltage regulators (DAVR) at asynchronous switching mode, when increasing the input supply voltage, a mode of simultaneous triggering of thyristors connected in series from two neighboring controllable switches is possible. This leads to arising of short circuit mode for the corresponding autotransformer section. The paper is dedicated to the study of the character and the parameters of this mode as a function of the input supply voltage magnitude and the load magnitudes. For this purpose, a vector analysis and the respective vector measurements of the quantities, describing the extreme work modes of DAVR is conducted.*

Keywords: vector analysis, vector measurement, asynchronous switch mode, thyristor, voltage regulator, semiconductor switch.

1. Introduction.

In Fig. 1 the replacement scheme of an autotransformer discrete alternating voltage regulator (ADAVR) with four thyristor semiconductor commutation elements (SCE) is shown [1, 2, 3]. When two adjacent semiconductor switches are switched unsynchronously and the input supply voltage increases, conditions are created for a possible "short circuit" mode for any of the windings w_1 and w_2 or w_3 (Fig.1).

The algorithm for commutation of adjacent SCE [4] includes:

- suspension of control pulses to the thyristors of the operating SCE (i.e. the thyristors in the SCE remain conducting until they are naturally switched);

- control pulses are applied to the SCE, which has to be switched on.

The comparative study should give an estimate of the load on the relevant winding and thyristors of the SCEs in such a mode.

Let us assume that thyristor commutator TC_4 is conducting current (i.e. thyristor T_{4a}, with TC_3, TC_2 and TC_1 do not do it) and due to an increase in the input supply voltage TS_4 should stop to conduct current and TC_3 should start to conduct. According to the adopted algorithm, this means that the switch S_4 has to be open first in an arbitrary moment. If the thyristor T_{4a} has been conducting within the half-period, it remains in this state regardless of the opening of switch S_4 at the moment of the corresponding switching.

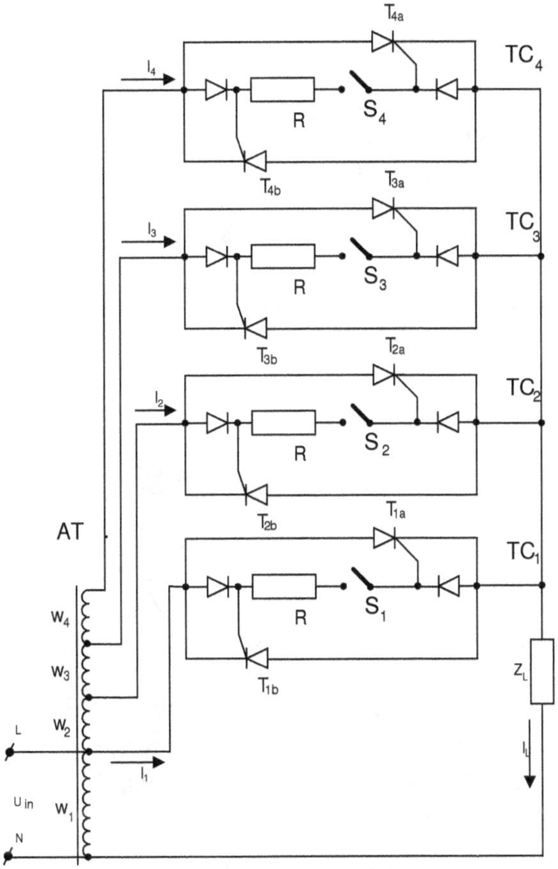

Fig. 1. Electrical scheme of ADAVR with four SCE.

After opening the switch S_4, the switch S_3 is closed (Fig. 1). This creates conditions for activation of the thyristor T_{3b}, if the following condition is fulfilled:

$$i_y = \frac{u_4 - u_d}{R} \geq i_{y\partial} \qquad (1)$$

where:

i_y is the current flowing through the controlling electrode of the corresponding thyristor;

$i_{y\partial}$ is a threshold value of the current i_y;

u_d is the voltage across the corresponding diode, which is conducting - Fig. 1;

u_4 is the instantaneous value of the voltage upon the fourth section of the autotransformer w_4;

When the thyristor T_{3b} starts to conduct for the time interval until the natural commutation of the thyristor T_{4a}, the fourth section of the winding of the autotransformer is short circuited by the conducting thyristors T_{4a} and T_{3b}, which are really connected in series.

If the switch S_3 is on, that happens in the time interval $0 \div t_1$ (Fig. 2), and then, T_{3b} will start to conduct with a delay after the moment t_1, because the inequality of equation (1) will be satisfied.

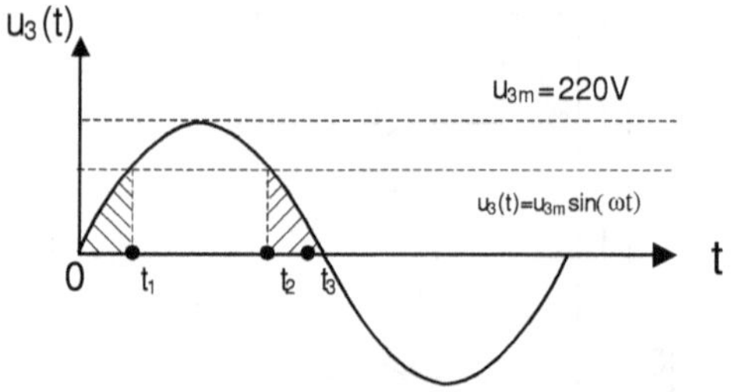

Fig. 2. Time diagram of the switching of thyristor K_3.

If switch S_3 is switched on in the interval $t_2 \div t_3$, the thyristor T_{3b} will not be able to start to conduct until the moment t_3, i.e. the short circuit mode for the fourth section of the winding will not occur.

The switch-on delay, mentioned so far at commutation angle φ in the interval $0 \div t_2$, is well illustrated in Fig. 3 ÷ 6, observed at load with parameters (*R_L=61,8Ω and L_L=1,76H - connected in parallel, at $\varphi=225^0$ and input voltage U_{IN}=160V, which is slowly increasing*). Here, the currents $i_{s3}(t)$ and $i_{s4}(t)$ flow through the switches S_3 and S_4 correspondingly.

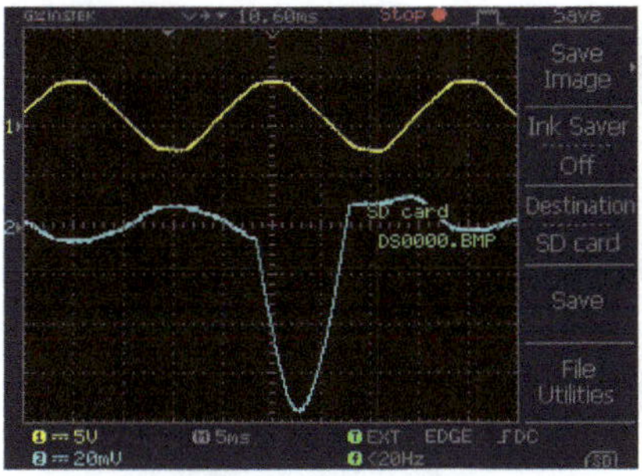

Fig. 3. *Oscillogram of* $i_1(t)$.

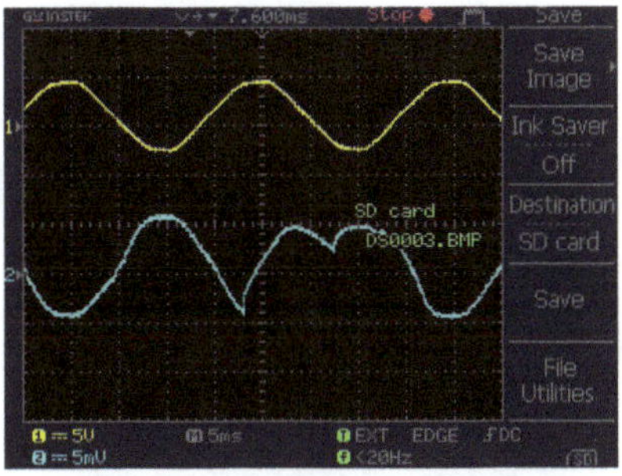

Fig. 4. *Oscillogram of* $i_2(t)$.

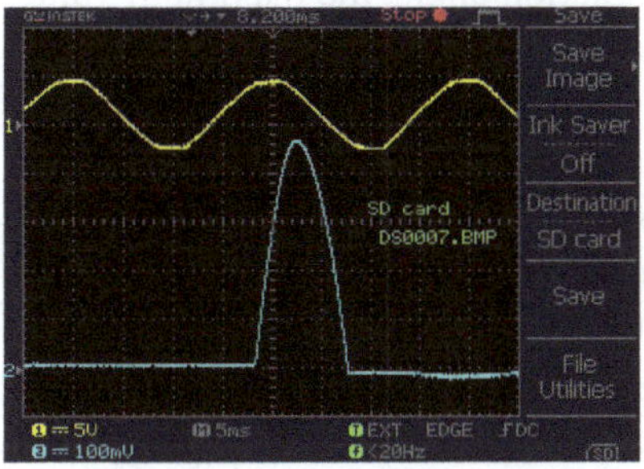

Fig. 5. *Oscillogram of* $i_{s3}(t)$.

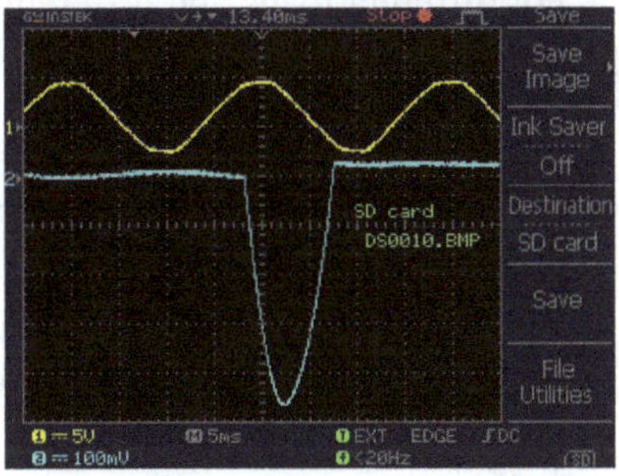

Fig. 6. *Oscillogram of* $i_{s4}(t)$.

The claim, that at a commutation angle in the range t2 ÷ t3 (Fig. 2) a short circuit for the respective windings w_2 and w_3 or w_4 does not occur, is illustrated in Fig. 7 ÷ 10. The oscillograms were taken at a switching angle $\varphi=315^0$ and $R_L=35.7\Omega$. On the oscillograms, the control pulse, supplied to the SCE, is exposed by ray № 2.

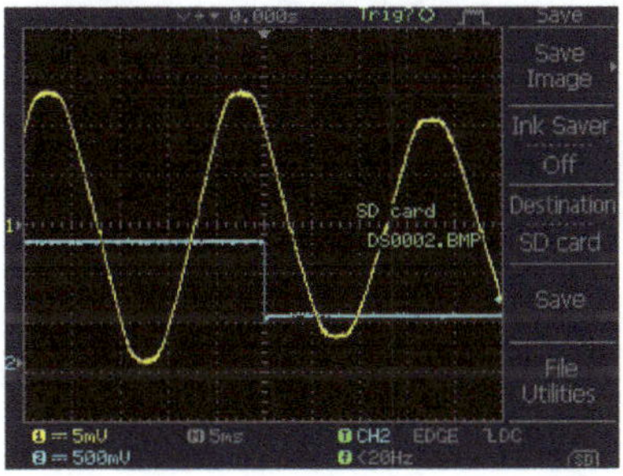

Fig. 7. *Oscillogram of* $i_1(t)$.

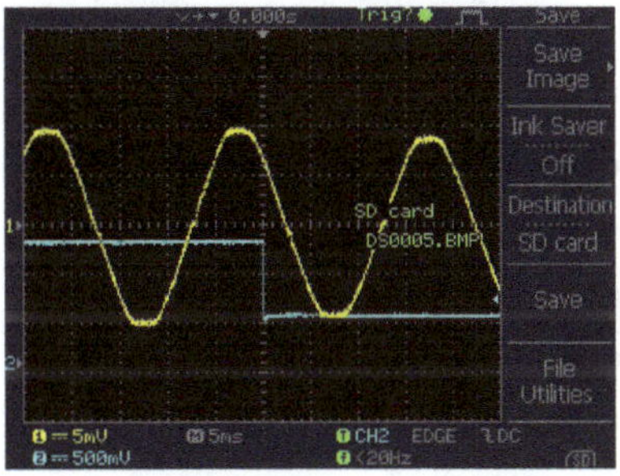

Fig. 8. *Oscillogram of* $i_2(t)$.

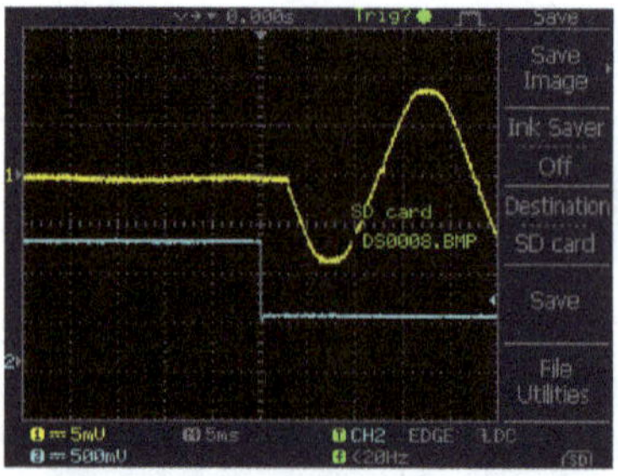

Fig. 9. *Oscillogram of* $i_{s3}(t)$.

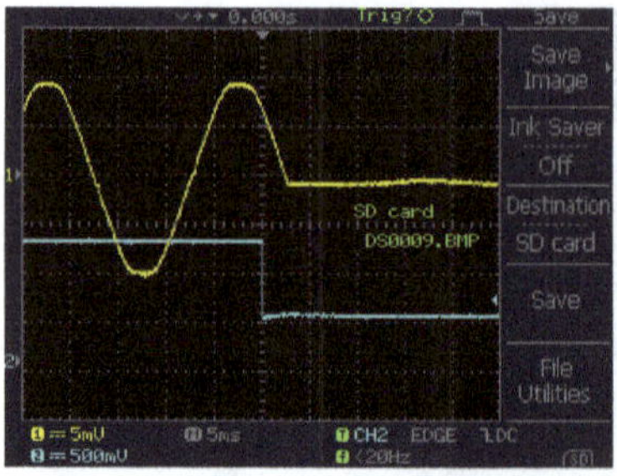

Fig. 10. *Oscillogram of* $i_{s4}(t)$.

For different loads and different commutation angle, the maximum values of the short circuit current and the duration of this mode are given in Table 1.

Table 1

Type of the load	**U_{in}=160V. Switch S_4 is off, and S_3 is on**		**U_{in}=180V. Switch S_3 is off, and S_2 is on**		**U_{in}=200V. Switch S_2 is off, and S_1 is on**	
	Maximum current	Time for commutation	Maximum current	Time for commutation	Maximum current	Time for commutation
	A	*ms*	*A*	*ms*	*A*	*mS*
Angle of commuta-tion	switch S_3	switch S_3	switch S_2	switch S_2	switch S_1	switch S_1
R – 225°	520	10	540	7,5	700	7
R – 270°	380	7,5	480	7	620	6,5
R*– **225°**	540	10	520	7,2	700	6,5
R* – **270°**	380	7,3	460	7	640	6
RL – 225°	500	10	520	7	680	8
RL – 270°	380	7,5	440	6	640	7
RC – 225°	540	10	520	8	800	8
RC– 270°	380	7,5	460	6,5	640	7

Note:

- R=61,52Ω;
- R*=35,7Ω;
- L=1,76H;
- C=38,08μF;
- RL are connected in parallel, RC are connected in series.

The load and switching duration specified in Table 1 for unsynchronized switching results in an overload, which in all cases is well below the maximum allowable for thyristors from the controlled thyristor switches (CTS).

Based on the models for computer simulation of the processes [5] for R, R-L and R-C loads, some oscillograms were obtained (Fig. 11a ÷ Fig. 14a), and these were compared with the computer simulations (Fig. 11b ÷ Fig. 14b). Unsynchronized switching is realized when the input supply voltage is increased for a load with parameters (R=35,7Ω, L=1,76H - connected in parallel) and commutation angle φ=225^0.

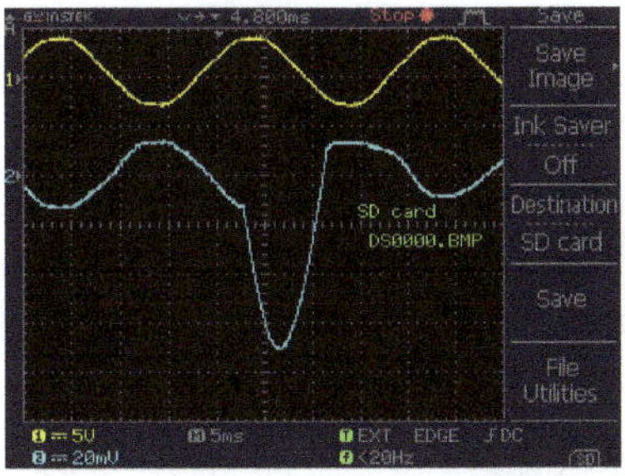

Fig. 11a. *Oscillogram of $i_1(t)$.*

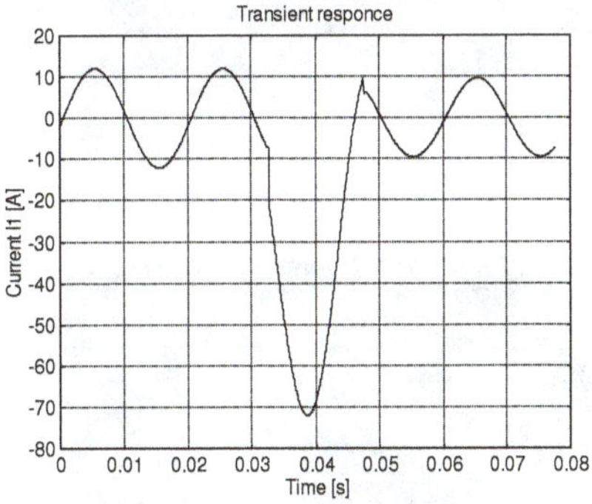

Fig. 11b. *Comp. simulation of $i_1(t)$.*

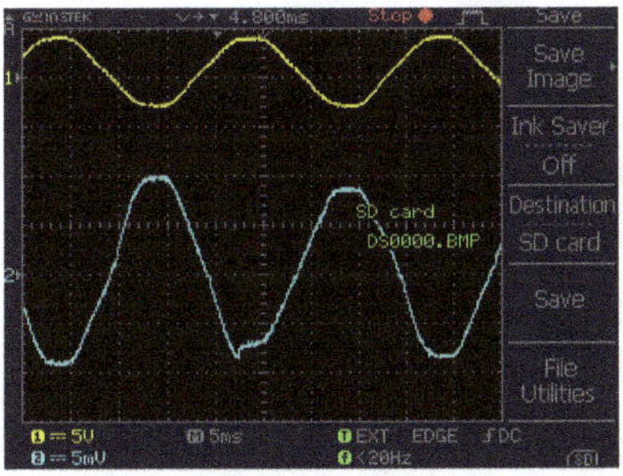

Fig. 12a. *Oscillogram of $i_2(t)$.*

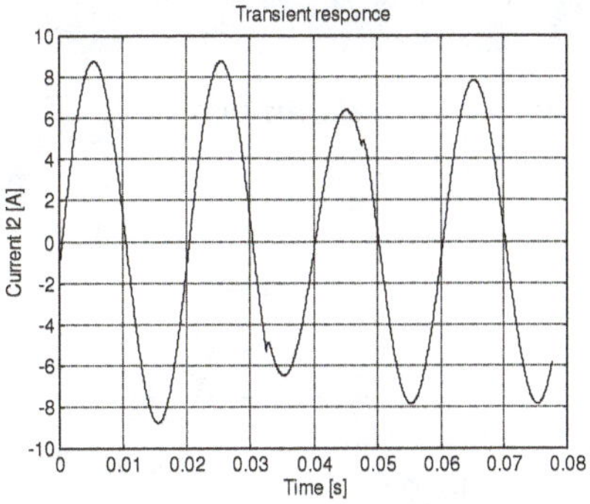

Fig. 12a. *Comp. simulation of $i_2(t)$.*

Table 2 presents the results of the vector measurements of currents and voltages of the ADAVR from Fig. 1, after the switching at U_{IN}=160V and at slow increase of the input supply voltage.

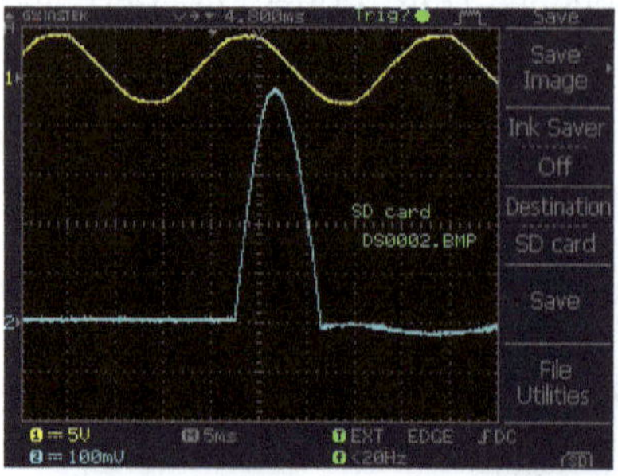

Fig. 13a. *Oscillogram of* $i_3(t)$.

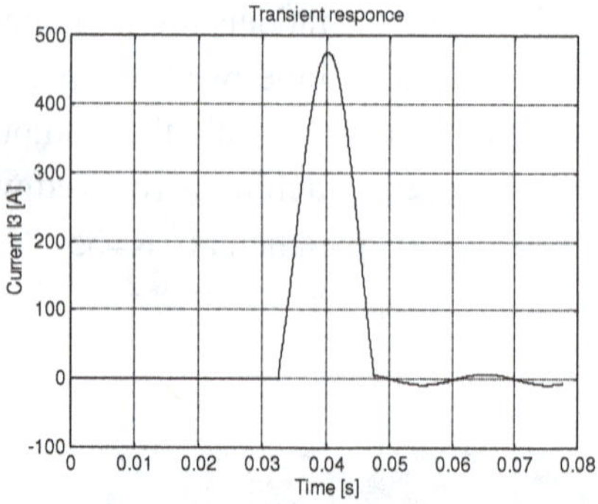

Fig. 13b. *Comp. simulation of* $i_3(t)$.

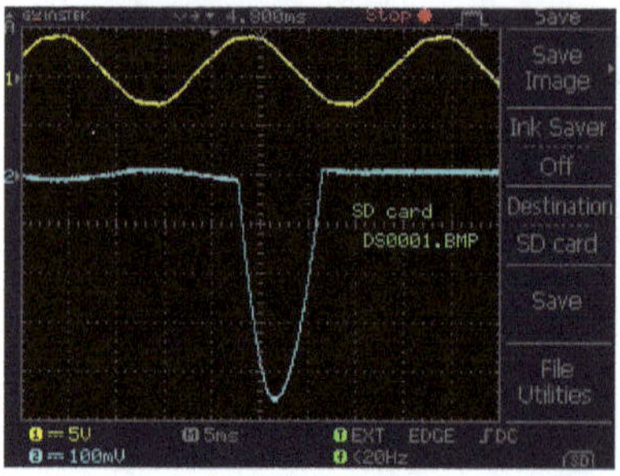

Fig. 14a. *Oscillogram of* $i_4(t)$.

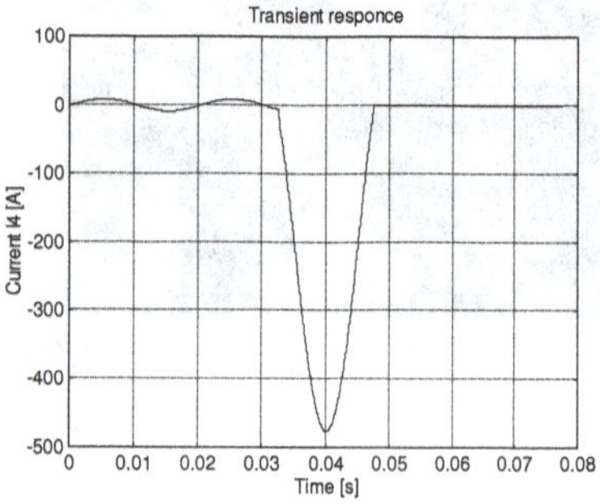

Fig. 14b. *Comp. simulation of* $i_4(t)$.

Table 2

Load	**Results from the vector measurements**		
	$\dot{I}_1$	$\dot{I}_2$	$\dot{U}_2$
	A	A	V
R_L=35,7Ω at angle of commutation φ=315^0	6,68.$e^{-j0,045}$	5,46.$e^{-j0,02}$	195,16.$e^{-j0,02}$
R_L=35,7Ω ; L_L=1,76H (in parallel) at angle of commutation φ=225^0	6,79.$e^{-j0,11}$	5,55.$e^{-j0,08}$	194,9.$e^{-j0,02}$

The scientific research, the results of which are presented in this publication, have been carried out under the project NP1/2013 within the inherent research activities of TU-Varna, funded by the state budget.

2. Conclusions.

1. The obtained results show that in the case of unsynchronized switching and increasing the magnitude of the input supply voltage, the emerging switching process implies overloading of the elements in the respective circuits.

2. The overload is significantly below the maximum permissible impact non-repetitive current for the thyristors of the SCE.

3. The proposed mathematical model allows to analyze and visualize the occurring specific short circuit mode in the case of unsynchronized switching of two adjacent semiconductor switches, when the input supply voltage increases, and to determine the magnitude and duration of the electrical load upon the autotransformer windings and the thyristors in the semiconductor switches.

4. The differences between the experimental and analytically obtained results do not exceed 10%, i.e. there is good coincidence.

References:

1. Barudov E., Panov E., Barudov S., Study of a precise non-linear model of an autotransformer discrete voltage regulator with semiconductor switching elements. Annual of TU-Varna, 2007, pp. 3÷9, ISSN 1311-896X.

2. Panov E., Barudov E., Barudov S., Advanced algorithm for analysis of a precise non-linear model of an autotransformer discrete voltage regulator with semiconductor switching elements. Annual of TU-Varna, 2009, Bulgaria, pp. 47÷52, ISSN: 1311-896X.

3. *Kularatna N.,* Power Sources and Supplies. Boca Raton FL, 2008.

4. Barudov E., Panov E., Barudov S., Comparative study of the switching processes in autotransformer discrete step-up voltage regulator. ELMA 2011, pp. 55÷60, ISSN 1313-4965.

5. Barudov E., Panov E., Barudov S., Analysis of Electrical Processes in Alternating Voltage Control Systems. Journal of International Scientific Publication: Materials, Methods & Technologies, 2010, Vol. 4, Part 1, pp. 154÷182, ISSN 1313-2539.

Chapter 5. Analysis of the Sensitivity of ADAVR

STUDY OF THE SENSITIVITY OF THE MODELS OF AUTOTRANSFORMER DISCRETE ALTERNATING VOLTAGE REGULATOR

Stefan T. Barudov*, Emil I. Panov**, Emil S. Barudov**

* Naval Academy "N. Y. Vaptsarov", Varna

** Technical University – Varna

(Published on the Marine Scientific Forum, Electronics, Electrical Engineering and Automatics. Informatics, VVMU "N. Y. Vaptsarov", Vol. 4, 2013, pp. 97-103, ISSN 1310-9278. (in Bulgarian))

Abstract: *The objects of the research are precise and approximate models of an autotransformer discrete alternating voltage regulator (ADAVR). The comparative evaluation is realized at different characters of the load – R, RL, RC and different commutation angles, for asynchronous switching modes. A vector analysis and subsequent vector measurements of the examined quantities, characterizing the electrical processes in ADAVR, are conducted. The obtained results give the possibility to select the parameters of the power circuits in the ADAVR and a choice of the thyristors in the semiconductor switches.*

Keywords: *vector analysis, vector measurement, model sensitivity, thyristor, voltage regulator, semiconductor switch.*

1. Introduction.

Fig. 1 shows the replacement scheme of the precise model of an autotransformer discrete alternating voltage regulator (ADAVR) with four thyristor semiconductor commutation elements (SCE) [1, 2, 3, 4]. The complete mathematical description of the exact active load model (for example, only for the II^{-nd} and III^{-rd} interval, i.e., the first and second transients) [5] is represented by a system of equations (1). If the losses in the magnetic core are ignored, the thyristors are assumed to be ideal switches, and the RC groups parallel to the thyristors do not exist, the replacement scheme takes the form of Fig. 2. The description of the approximate model for the active load is represented by a system of equations (2).

Based on the assumptions adopted, the dimensionality of the systems of equations is reduced, easing the process of mathematical analysis. The aim of the present work is to conduct a comparative evaluation based on analytical and experimental results for specific prototypes.

2. Presentation.

Table 1 presents the data from the simulations and the experimental study of a particular ADAVR, (at the end of the third interval at steady-state regime for different loads) and Table 2 presents the percentage deviation of the simulation results from the experimental data.

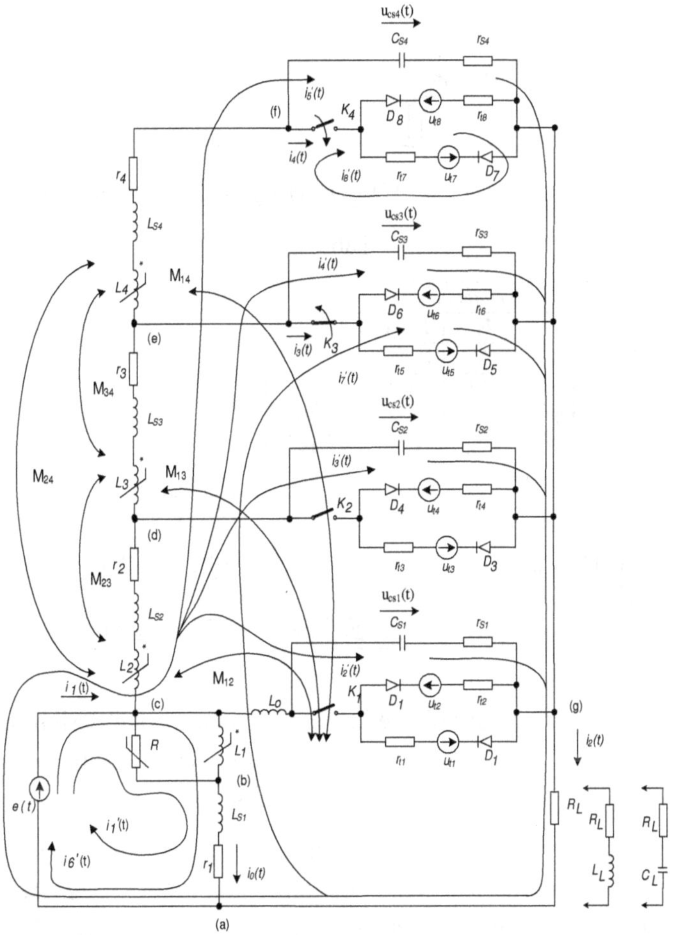

Fig. 1. *Replacement scheme of ADAVR with four SCE (precise model).*

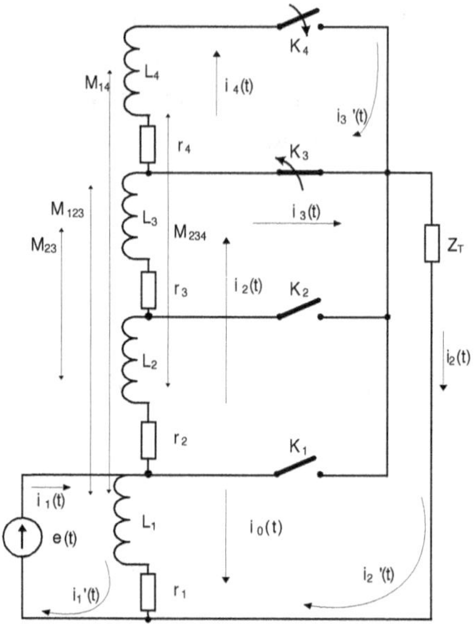

Fig. 2. *Replacement scheme of ADAVR with four SCE (approximate model).*

Table 1

	Experimental data				
R_T	**U_1**	**I_1**	**I_0**	**$I_2 \equiv I_T$**	**$U_2 \equiv U_T$**
Ω	**V**	**A**	**A**	**A**	**V**
115,8	160	2,69	0,81	1,9	220
61,8	160	4,94	1,42	3,56	220
35,5	160	8,45	2,38	6,16	219

Precise simulations					*Approximate simulations*				
U_1	**I_1**	**I_0**	**$I_2 \equiv I_T$**	**$U_2 \equiv U_T$**	**U_1**	**I_1**	**I_0**	**$I_2 \equiv I_T$**	**$U_2 \equiv U_T$**
V	**A**	**A**	**A**	**V**	**V**	**A**	**A**	**A**	**V**
160	2,6722	0,79369	1,8908	218,94	160	2,5893	0,71111	1,8924	219,12
160	4,9303	1,3977	3,5397	218,74	160	4,7494	1,2981	3,4513	213,28
160	8,4942	2,3582	6,1401	218,29	160	8,3958	2,2768	6,1235	218,70

The parameter output voltage U_2 is basic one for the design of ADAVR. The average error δ for U_2 of the approximate model at the presented loads is δ=1,3%, and that of the precise model is δ=0,83%.

In Fig. 3a, b, c ÷ Fig. 6a, b, c the oscillograms from the experiments and the computer simulations at R-L load are presented *(R_L=35,7Ω and L_L=1,76H connected in parallel, angle of commutation φ=270^0 and U_{IN}=160V, which is slowly decreasing).*

Table 2

	Precise model			*Approximate model*		
R_T	**I_1**	**$I_2 \equiv I_T$**	**U_2**	**I_1**	**$I_2 \equiv I_T$**	**U_2**
Ω	%	%	%	%	%	%
R load						
115,8	-0,66	-0,48	-0,48	-3,74	-0,4	-0,4
61,8	-0,196	-0,57	-0,57	-3,86	-3,05	-3,05
35,5	+0,52	-0,32	-0,32	-0,64	-0,59	-0,14
R-L load, L=1,0460H						
115,8	-5,57	-2,84	-0,76	-11,62	-2,86	-0,73
61,8	-3,55	+2,18	-0,72	-7,27	+1,94	-0,73
35,5	-11,71	-11,56	-0,89	-13,22	-11,69	-0,86
R-C load, C=38,09μF						
115,8	-0,37	-1,01	+0,09	-7,16	-0,37	+0,74
61,8	+1,23	+0,93	+2,38	-4,99	+0,07	+1,51
35,5	-1,42	-5,28	+1,34	-4,53	-4,98	+1,73

If we tentatively label the compared models as "precise" and "approximate" as indicated in Table 2, there are more significant deviations in the approximate model for practically all loads.

A system of equations for a precise model of the ADAVR (II^{-nd} interval i.e. first transient) at R load. The system of equations is described by the method with state variables.

$$+L_{S1}.\frac{di_1^{'}(t)}{dt}+0.\frac{di_2^{'}(t)}{dt}+0.\frac{di_3^{'}(t)}{dt}+0.\frac{di_4^{'}(t)}{dt}+0.\frac{di_5^{'}(t)}{dt}+L_{S1}.\frac{di_6^{'}(t)}{dt}-$$
$$-L_{S1}.\frac{di_7^{'}(t)}{dt}+0.\frac{di_8^{'}(t)}{dt}=-(r_1+R_{\mu}).i_1^{'}(t)+0.i_2^{'}(t)+0.i_3^{'}(t)+0.i_4^{'}(t)+$$
$$+0.i_5^{'}(t)-r_1.i_6^{'}(t)+r_1.i_7^{'}(t)+0.i_8^{'}(t)+e(t)$$

$$+0.\frac{di_1^{'}(t)}{dt}+L_0.\frac{di_2^{'}(t)}{dt}+0.\frac{di_3^{'}(t)}{dt}+0.\frac{di_4^{'}(t)}{dt}+0.\frac{di_5^{'}(t)}{dt}+0.\frac{di_6^{'}(t)}{dt}+$$
$$+0.\frac{di_7^{'}(t)}{dt}+0.\frac{di_8^{'}(t)}{dt}=0.i_1^{'}(t)-(r_{S1}+R_T).i_2^{'}(t)-R_T.i_3^{'}(t)-R_T.i_4^{'}(t)-$$
$$-R_T.i_5^{'}(t)+0.i_6^{'}(t)-R_T.i_7^{'}(t)+0.i_8^{'}(t)-u_{CS1}+e(t)$$

$$+0.\frac{di_1'(t)}{dt}+0.\frac{di_2'(t)}{dt}+(L_2+L_{S2}).\frac{di_3'(t)}{dt}+(L_2+L_{S2}+M_{23}).\frac{di_4'(t)}{dt}+$$

$$+(L_2+L_{S2}+M_{23}+M_{24}).\frac{di_5'(t)}{dt}-M_{12}.\frac{di_6'(t)}{dt}+(L_2+L_{S2}+M_{12}+M_{23}).\frac{di_8'(t)}{dt}+$$

$$+M_{24}.\frac{di_8'(t)}{dt}=0.i_1'(t)-R_T.i_2'(t)-(r_2+r_{S2}+R_T).i_3'(t)-(r_2+R_T).i_4'(t)-$$

$$-(r_2+R_T).i_5'(t)+0.i_6'(t)-(r_2+R_T).i_7'(t)+0.i_8'(t)-u_{CS2}(t)+e(t)$$

$$+0.\frac{di_1'(t)}{dt}+0.\frac{di_2'(t)}{dt}+(L_2+L_{S2}+M_{23}).\frac{di_3'(t)}{dt}+(L_2+L_{S2}+L_3+L_{S3}+2.M_{23}).\frac{di_4'(t)}{dt}+$$

$$+(L_2+L_{S2}+L_3+L_{s3}+2.M_{23}+M_{24}).\frac{di_5'(t)}{dt}-(M_{12}+M_{13}).\frac{di_6'(t)}{dt}+(L_2+L_{S2}+L_3+L_{S3}+$$

$$+2.M_{23}+M_{12}+M_{13}).\frac{di_8'(t)}{dt}=0.i_1'(t)-R_T.i_2'(t)-(r_2+R_T).i_3'(t)-(r_2+r_3+r_{S3}+R_T).i_4'(t)-$$

$$-(r_2+r_3+R_T).i_5'(t)+0.i_6'(t)-(r_2+r_3+R_T).i_7'(t)+0.i_8'(t)-u_{CS3}(t)+e(t)$$

$$+0.\frac{di_1'(t)}{dt}+0.\frac{di_2'(t)}{dt}+(L_2+L_{S2}+M_{23}+M_{24}).\frac{di_3'(t)}{dt}+(L_2+L_{S2}+L_3+L_{S3}+$$

$$+2.M_{23}+M_{24}+M_{34}).\frac{di_4'(t)}{dt}+(L_2+L_{S2}+L_3+L_{S3}+L_4+L_{S4}+2.M_{23}+2.M_{34}+$$

$$+2.M_{24}).\frac{di_5'(t)}{dt}-(M_{12}+M_{13}+M_{14}).\frac{di_6'(t)}{dt}+(L_2+L_{S2}+L_3+L_{S3}+2.M_{23}+M_{24}+$$

$$+M_{34}+M_{12}+M_{13}+M_{14}).\frac{di_7'(t)}{dt}+(L_4+L_{S4}+M_{23}+M_{34}).\frac{di_8'(t)}{dt}=0.i_1'(t)-R_T.i_2'(t)-$$

$$-(r_2+R_T).i_3'(t)-(r_2+r_3+R_T).i_4'(t)-(r_2+r_3+r_4+r_{S4}+R_T).i_5'(t)+0.i_6'(t)-$$

$$-(r_2+r_3+R_T).i_7'(t)-r_4.i_8'(t)-u_{CS4}(t)+e(t)$$

$$-L_{S1}.\frac{di_1'(t)}{dt}+0.\frac{di_2'(t)}{dt}+(L_2+L_{S2}+M_{12}+M_{23}).\frac{di_3'(t)}{dt}+(L_2+L_{S2}+L_3+L_{S3}+2.M_{23}+$$

$$+M_{12}+M_{23}).\frac{di_4'(t)}{dt}+(L_2+L_{S2}+L_3+L_{S3}+2.M_{23}+M_{12}+M_{13}+M_{14}+M_{24}+M_{34}).\frac{di_5'(t)}{dt}-$$

$$-(L_1+L_{S1}+M_{12}+M_{13}).\frac{di_6'(t)}{dt}+(L_1+L_{S1}+L_2+L_{S2}+L_3+L_{S3}+2.M_{12}+2.M_{13}+2.M_{23}).\frac{di_7'(t)}{dt}+$$

$$+(M_{14}+M_{22}+M\,1_{23}).\frac{di_8'(t)}{dt}=+r_1.i_1'(t)-R_T.i_2'(t)-(r_2+R_T).i_3'(t)-(r_2+r_3+R_T).i_4'(t)-$$

$$-(r_2+r_3+R_T).i_5'(t)+r_1.i_6'(t)-(r_1+r_2+r_3+r_{t6(5)}+R_T).i_7'+r_{t6(5)}.i_8'(t)-sign\left[i_7'(t)\right].u_{t6}$$

$$+0.\frac{di_1'(t)}{dt}+0.\frac{di_2'(t)}{dt}+M_{24}.\frac{di_3'(t)}{dt}+(M_{22}+M_{34}).\frac{di_4'(t)}{dt}+(L_4+L_{S4}+M_{24}+M_{34}).\frac{di_5'(t)}{dt}-$$

$$-M_{14}.\frac{di_6'(t)}{dt}+(M_{14}+M_{24}+M_{34}).\frac{di_7'(t)}{dt}+(L_4+L_{S4}).\frac{di_8'(t)}{dt}=0.i_1'(t)+0.i_2'(t)+0.i_3'(t)+$$

$$+0.i_4'(t)-r_4.i_5'(t)+0.i_6'(t)+r_{t6(5)}.i_7'(t)-(r_4+r_{t6(5)}+r_{t8(7)}).i_8'(t)-sign\left[i_8'(t)\right].u_{t8}+sign.\left[i_7'(t)\right].u_{t6}$$

$$C_{S1}.\frac{du_{CS1}(t)}{dt}=i_2'(t)$$

$$C_{S2}.\frac{du_{CS2}(t)}{dt}=i_3'(t)$$

$$C_{S3}.\frac{du_{CS3}(t)}{dt}=i_4'(t)$$

$$C_{S4}.\frac{du_{CS4}(t)}{dt}=i_5'(t)$$

A system of equations for a precise model of the ADAVR (III^{-rd} interval i.e. second transient) at R load. The system of equations is described by the method with state variables.

$$+L_{S1}.\frac{di_1'(t)}{dt}+0.\frac{di_2'(t)}{dt}+0.\frac{di_3'(t)}{dt}+0.\frac{di_4'(t)}{dt}+0.\frac{di_5'(t)}{dt}+L_{S1}.\frac{di_6'(t)}{dt}-L_{S1}.\frac{di_8'(t)}{dt}=$$

$$=-(r_1+r_\mu).i_1'(t)+0.i_2'(t)+0.i_3'(t)+0.i_4'(t)+0.i_5'(t)-r_1.i_6'(t)+r_1.i_8'(t)+e(t)$$

$$+0.\frac{di_1'(t)}{dt}+L_0.\frac{di_2'(t)}{dt}+0.\frac{di_3'(t)}{dt}+0.\frac{di_4'(t)}{dt}+0.\frac{di_5'(t)}{dt}+0.\frac{di_6'(t)}{dt}+0.\frac{di_8'(t)}{dt}=$$

$$=0.i_1'(t)-(r_{S1}+R_T).i_2'(t)-R_T.i_3'(t)-R_T.i_4'(t)-R_T.i_5'(t)+0.i_6'(t)-R_T.i_8'(t)-u_{CS1}+e(t)$$

$$+0.\frac{di_1'(t)}{dt}+0.\frac{di_2'(t)}{dt}+(L_2+L_{S2}).\frac{di_3'(t)}{dt}+(L_2+L_{S2}+M_{23}).\frac{di_4'(t)}{dt}+$$

$$+(L_2+L_{S2}+M_{23}+M_{24}).\frac{di_5'(t)}{dt}-M_{12}.\frac{di_6'(t)}{dt}+(L_2+L_{S2}+M_{12}+M_{23}+M_{24}).\frac{di_8'(t)}{dt}=$$

$$0.i_1'(t)-R_T.i_2'(t)-(r_2+r_{S2}+R_T).i_3'(t)-(r_2+R_T).i_4'(t)-(r_2+R_T).i_5'(t)+0.i_6'(t)-$$

$$-(r_2+R_T).i_8'(t)-u_{CS2}(t)+e(t)$$

$$+0.\frac{di_1'(t)}{dt}+0.\frac{di_2'(t)}{dt}+(L_2+L_{S2}+M_{23}).\frac{di_3'(t)}{dt}+(L_2+L_{S2}+L_3+L_{S3}+2.M_{23}).\frac{di_4'(t)}{dt}+$$
$$+(L_2+L_{S2}+L_3+L_{s3}+2.M_{23}+M_3+M_{24}).\frac{di_5'(t)}{dt}-(M_{12}+M_{13}).\frac{di_6'(t)}{dt}+$$
$$+(L_2+L_{S2}+L_3+L_{S3}+2.M_{23}+M_{12}+M_{13}+M_{24}+M_{34}).\frac{di_8'(t)}{dt}=0.i_1'(t)-R_T.i_2'(t)-$$
$$-(r_2+R_T).i_3'(t)-(r_2+r_3+r_{S3}+R_T).i_4'(t)-(r_2+r_3+R_T).i_5'(t)+0.i_6'(t)-(r_2+r_3+R_T).i_8'(t)-$$
$$-u_{CS3}(t)+e(t)$$

$$+0.\frac{di_1'(t)}{dt}+0.\frac{di_2'(t)}{dt}+(L_2+L_{S2}+M_{23}+M_{24}).\frac{di_3'(t)}{dt}+(L_2+L_{S2}+L_3+L_{S3}+$$
$$+2.M_{23}+M_{24}+M_{34}).\frac{di_4'(t)}{dt}+(L_2+L_{S2}+L_3+L_{S3}+L_4+L_{S4}+2.M_{23}+2.M_{24}+2.M_{34}).\frac{di_5'(t)}{dt}-$$
$$-(M_{12}+M_{13}+M_{14}).\frac{di_6'(t)}{dt}+(L_2+L_{S2}+L_3+L_{S3}+2.M_{23}+2.M_{24}+2.M_{34}+M_{12}+M_{13}+$$
$$+M_{14}+L_4+L_{S4}).\frac{di_8'(t)}{dt}=0.i_1'(t)-R_T.i_2'(t)-(r_2+R_T).i_3'(t)-(r_2+r_3+R_T).i_4'(t)-(r_2+r_3+r_4+$$
$$+r_{S4}+R_T).i_5'(t)+0.i_6'(t)-(r_2+r_3+r_4+R_T).i_8'(t)-u_{CS4}(t)+e(t)$$

$$+L_{S1}.\frac{di_1'(t)}{dt}+0.\frac{di_2'(t)}{dt}-M_{12}.\frac{di_3'(t)}{dt}-(M_{12}+M_{13}).\frac{di_4'(t)}{dt}-(M_{12}+M_{13}+M_{14}).\frac{di_5'(t)}{dt}+$$
$$+(L_1+L_{S1}).\frac{di_6'(t)}{dt}-(L_1+L_{S1}+M_{12}+M_{13}+M_{14}).\frac{di_8'(t)}{dt}=-r_1.i_1'(t)+0.i_2'(t)+0.i_3'(t)+$$
$$+0.i_4'(t)+0.i_5'(t)-r_1.i_6'(t)+r_1.i_8'(t)+e(t)$$

$$-L_{S1}.\frac{di_1'(t)}{dt}+0.\frac{di_2'(t)}{dt}+(L_2+L_{S2}+M_{12}+M_{23}+M_{24}).\frac{di_3'(t)}{dt}+(L_2+L_{S2}+L_3+L_{S3}+$$
$$+2.M_{23}+M_{12}+M_{24}+M_{34}).\frac{di_4'(t)}{dt}+(L_2+L_{S2}+L_3+L_{S3}+L_4+L_{S4}+2.M_{23}+M_{12}+$$
$$+M_{13}+M_{14}+2.M_{24}+2.M_{34}).\frac{di_5'(t)}{dt}-(L_1+L_{S1}+M_{12}+M_{13}+M_{14}).\frac{di_6'(t)}{dt}+$$
$$+(L_1+L_{S1}+L_2+L_{S2}+L_{S3}+2.M_{12}+2.M_{13}+2.M_{23}+L_4+L_{S4}+2.M_{14}+2.M_{24}+$$
$$+2.M_{34}).\frac{di_8'(t)}{dt}=+r_1.i_1'(t)-R_T.i_2'(t)-(r_2+R_T).i_3'(t)-(r_2+r_3+R_T).i_4'(t)-(r_2+r_3+r_4+$$
$$+R_T).i_5'(t)+r_1.i_6'(t)-(r_1+r_2+r_3+r_4+r_{t7(8)}+R_T).i_8'-sign\left[i_8'(t)\right].u_{t8(7)}$$

$$
\begin{aligned}
&C_{S1}.\frac{du_{cs1}(t)}{dt}=i_2'(t)\\
&C_{S2}.\frac{du_{cs2}(t)}{dt}=i_3'(t)\\
&C_{S3}.\frac{du_{cs3}(t)}{dt}=i_4'(t)\\
&C_{S4}.\frac{du_{cs4}(t)}{dt}=i_5'(t)
\end{aligned}
\qquad (1)
$$

System of equations for an approximate model of the ADAVR at R loads in I^{-st}, II^{-nd} and III^{-rd} intervals.

$$
\begin{cases}
(r_1+j\omega L_1).\dot{I}_1'-\left[r_1+j\omega\left(L_1+M_{123}\right)\right].\dot{I}_2'=\dot{E}\\
-\left[r_1+j\omega\left(L_1+M_{123}\right)\right].\dot{I}_1'+\left[\left(R_T+r_1+r_2+r_3\right)+j\omega\left(L_1+2M_{123}+L_{23}\right)\right]\dot{I}_2'=0
\end{cases}
$$

$$
\begin{cases}
L_1.\frac{di_1'(t)}{dt}-\left(L_1+M_{123}\right)\frac{di_2'(t)}{dt}-M_{14}.\frac{di_3'(t)}{dt}=-r_1.i_1'(t)+r_1.i_2'(t)+e_m\sin(\omega t+\phi)\\
-\left(L_1+M_{123}\right)\frac{di_1'(t)}{dt}+\left(L_1+2M_{123}+L_{23}\right)\frac{di_2'(t)}{dt}+\left(M_{14}+M_{234}\right)\frac{di_3'(t)}{dt}=\\
=-r_1.i_1'(t)-\left(R_T+r_1+r_2+r_3\right).i_2'(t)\\
-M_{14}\frac{di_1'(t)}{dt}+\left(M_{14}+M_{234}\right)\frac{di_2'(t)}{dt}+L_4\frac{di_3'(t)}{dt}=-r_4.i_3'(t)
\end{cases}
$$

(2)

$$
\begin{cases}
L_1.\frac{di_1'(t)}{dt}-\left(L_1+M_{1234}\right)\frac{di_2'(t)}{dt}=-r_1.i_1'(t)+r_1.i_2'(t)+e_m\sin(\omega t+\phi)\\
-\left(L_1+M_{1234}\right)\frac{di_1'(t)}{dt}+\left(L_1+2M_{1234}+L_{234}\right)\frac{di_2'(t)}{dt}=-r_1.i_1'(t)-\left(R_T+r_1+r_2+r_3+r_4\right).i_2'(t)
\end{cases}
$$

Graphs - comparison between experiments and simulations for R-L load.

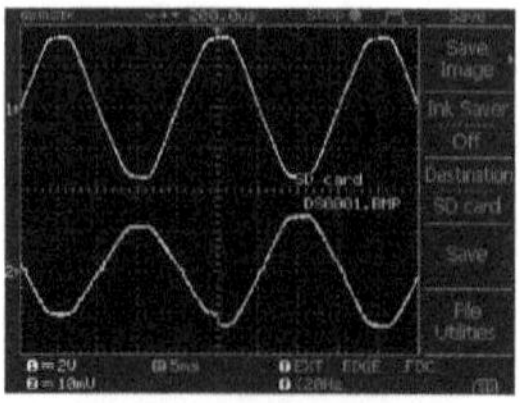

Fig. 3.a. *Oscillogram of* $i_1(t)$.

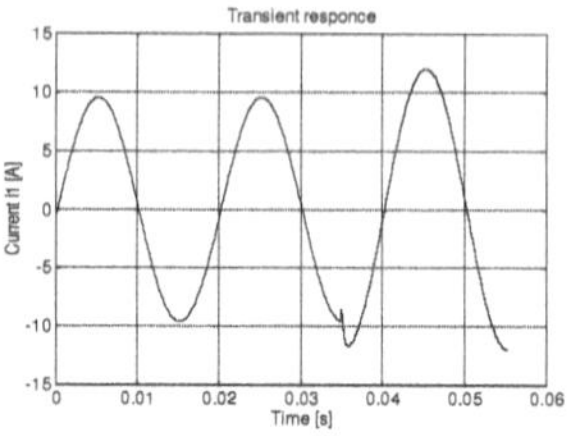

Fig. 3.b. *Simulation of* $i_1(t)$ *"precise" model.*

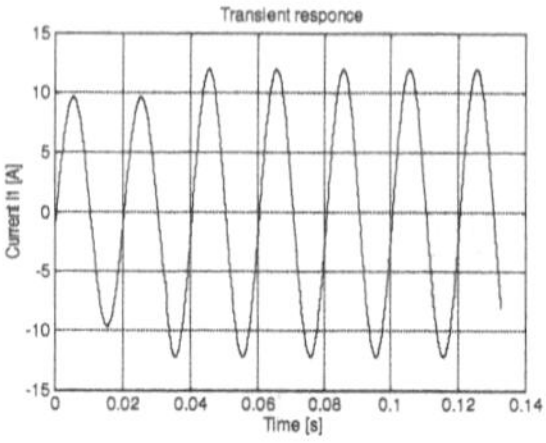

Fig. 3.c. *Simulation of* $i_1(t)$ *„approximate" model.*

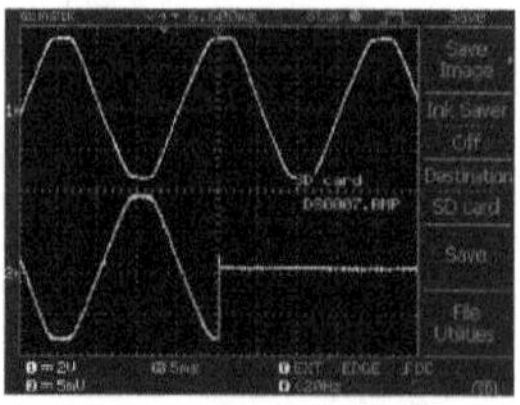

Fig. 4.a. *Oscillogram of* $i_{s3}(t)$.

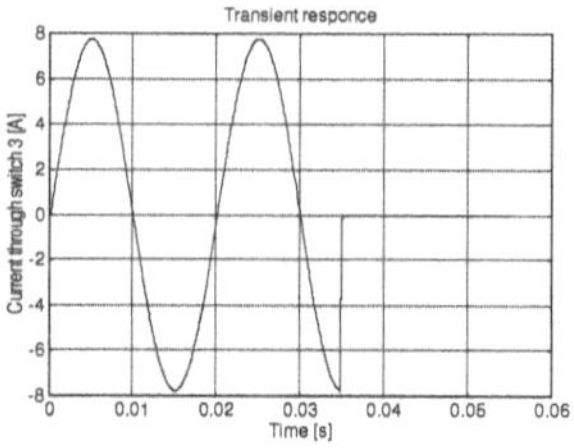

Fig. 4.b. *Simulation of* $i_{s3}(t)$ *"precise" model.*

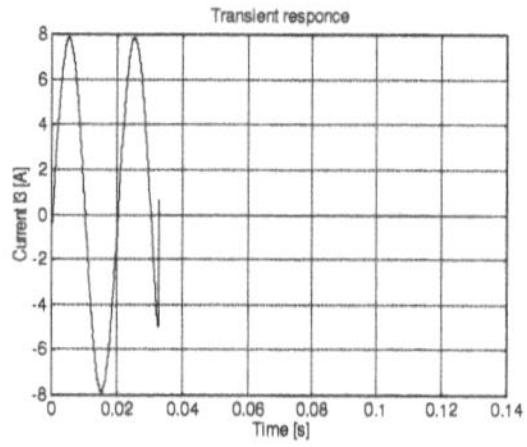

Fig. 4.c. *Simulation of* $i_{s3}(t)$ $i_{s3}(t)$ *approximate" model.*

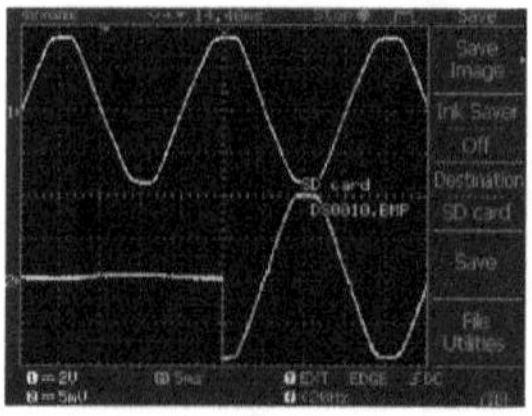

Fig. 5.a. *Oscillogram of* $i_{s4}(t)$.

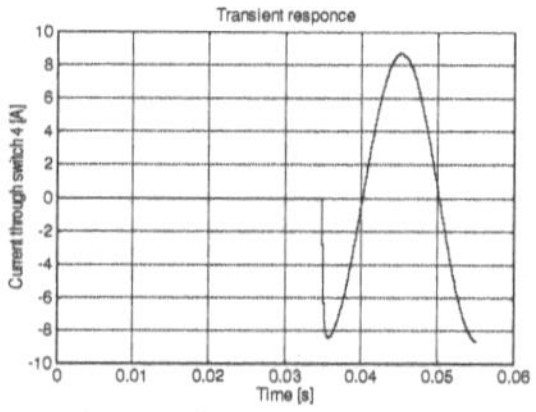

Fig. 5.b. *Simulation of* $i_{s4}(t)$ *"precise" model.*

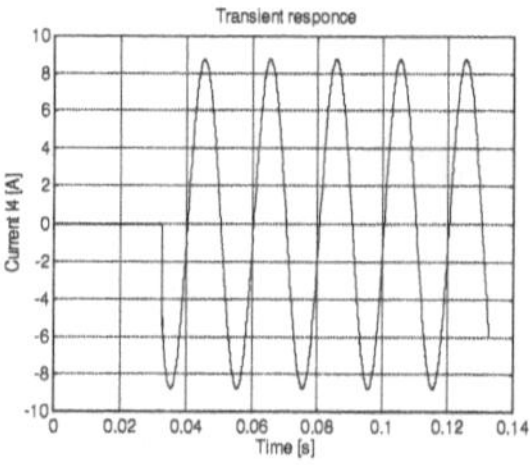

Fig. 5.c. *Simulation of* $i_{s4}(t)$ *„approximate"model.*

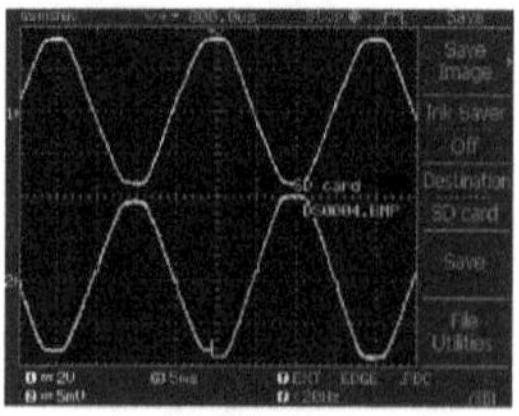

Fig. 6.a. *Oscillogram of* $i_2(t)$.

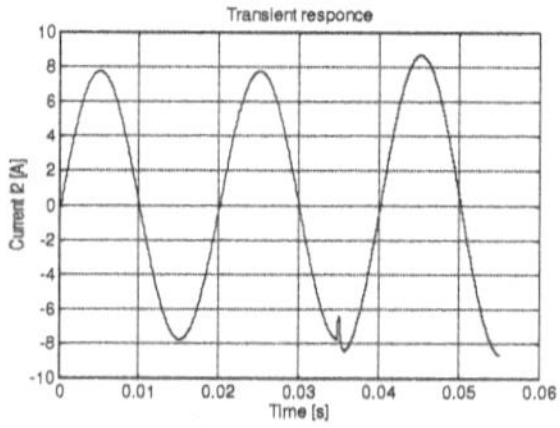

Fig. 6.b. *Simulation of* $i_2(t)$ *"precise" model.*

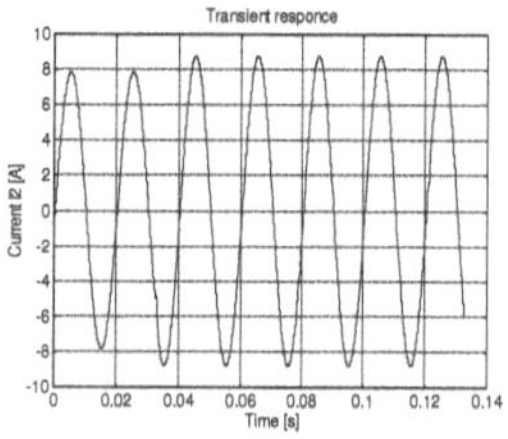

Fig. 6.c. *Simulation of* $i_2(t)$ *„approximate"model.*

It is clear, that the computer visualizations for the precision model of the switching process more accurately describe the oscillograms from the experimental study of the particular prototype.

The approximate model can be used for engineering calculations as the relative error δ in the determination of specific parameters rarely exceeds 1,3%, reaching 5% in single cases. In this model, the mathematical operations have been greatly simplified, which allows quick calculation of the quantities required for the design of the ADAVR in the MATLAB environment.

Table 3 shows the vector magnitudes of the ADAVR currents and voltages at some specific loads.

Table 3

Load	***Results from the vector measurements***		
	$\dot{I}_1$	$\dot{I}_2$	$\dot{U}_2$
	A	A	V
R_L=35,7Ω	$8{,}45.e^{-j0{,}08}$	$6{,}16.e^{-j0{,}06}$	$219.e^{-j0{,}06}$
R_L=35,7Ω; L_L=1,76H	$9{,}75.e^{-j0{,}14}$	$7{,}03.e^{-j0{,}12}$	$219{,}9.e^{j0{,}06}$
R_L=61,78Ω; C_L=38,08μF	$2{,}85.e^{j0{,}89}$	$2{,}15.e^{j0{,}92}$	$219.e^{-j0{,}01}$

The scientific research, the results of which are presented in this publication, have been carried out under the project NP1/2013 within the inherent research activities of TU-Varna, funded by the state budget.

3. Conclusions.

1. The adopted assumptions lead to a significant reduction in the dimensionality of the system equations describing the electrical equilibrium of the ADAVR, and thus facilitate the analytical study.
2. The obtained results indicate low sensitivity to the assumptions made, with errors in the analytical determination of specific parameters rarely exceeding 1,3%, and in single cases reaching 5% in the approximate model.
3. The inaccurate model can be used with sufficient accuracy for rapid engineering calculations of the ADAVR.

References:

1. *Barudov E., Barudov S., Panov E.*, Switching Processes in a Step Voltage Regulator. Acta Universitatis Pontica Euxinus – Volume IV, Number 1, 2005, pp. 21÷25, ISSN 1312-1669.
2. *Kularatna N.*, Power Sources and Supplies. Boca Raton FL, 2008.
3. *Harlow James H.*, Transformers. The Electric Power Engineering Handbook. Ed. L.L. Grigsby Boca Raton: CRC Press LLC, 2001.
4. *Lowdon E.*, Practical Transformer Design Handbook. Howard W. Sams Company, Indianapolis, III edition, 2005.
5. *Barudov E., Panov E., Barudov S.*, Analysis of Electrical Processes in Alternating Voltage Control Systems. Journal of International Scientific Publication: Materials, Methods & Technologies, 2010, Vol. 4, Part 1, pp. 154÷182, ISSN 1313-2539.

Chapter 6. Errors Between the Results of the Simulations and Physical Experiments

ERROR ANALYSIS OF THE VECTOR ANALALYSIS AND THE VECTOR MEASUREMENTS IN AN AUTOTRANSFORMER DISCRETE AC VOLTAGE REGULATOR

Emil Panov, Emil Barudov, Stefan Barudov

(Published on the Jubilee Scientific Conference "50 Years of ETET Department", 4-5 October 2013, Varna, Bulgaria, Annual of the Technical University - Varna, Vol. I, 2013, pp. 180-184, ISSN: 1311-896X. (in Bulgarian))

Abstract: The autotransformer discrete alternating voltage regulators (ADAVR) are electronic converters, which are situated between the transmission line (TL) and the load (L). They function in steady-state and commutation regimes. In their power circuits some non-linear elements are incorporated and their design is connected with the multi-factor analysis of the electric processes in the system "TL-ADAVR-L". The utilization of the results from the computer simulation analysis demands estimation of their plausibility. The paper is dedicated to the comparative exploration of the experimental and the theoretical results, which were received from different mathematical models, describing the system.
Keywords: discrete voltage regulator, thyristor, mathematical model, vector measurements, error analysis.

I. INTRODUCTION

The computer simulations shorten the time and the resources for sizing ADAVR - for solving various application problems. An important element in this process is the assessment of the reliability of the results of the computer simulations. Two mathematical models of the system "TL-ADAVR-L" for analysis of the transients in switching of controlled thyristor switches (CTS) have been developed - a precise model [1] and, assuming certain idealizations, an approximate model [2].

The paper is devoted to the comparative analysis of the results obtained using the two models and the experimental data from the corresponding vector measurements to assess the reliability of the models.

II. ANALYSIS

The relative absolute errors of the measurements of the quantities I_1, $I_1 \equiv I_T$ and U_2 (Fig. 1 and Fig. 2) are investigated using the formula:

$$Y \equiv \delta_x = \left| \frac{X_{measured} - X_{calculated}}{X_{measured}} \right| .100[\%] , \qquad (1)$$

where the quantity X may be I_1, I_2 or U_2.

The study was conducted with respect to the analysis of the obtained relative absolute errors δ_X in the determination of the quantities I_1, I_2, and U_2 for the precise and the approximate models for three types of R-, RL-, and RC- loads.

Fig. 1 shows the replacement scheme of the precise model of an autotransformer discrete ac voltage regulator with four controllable thyristor switches, and Fig. 2 shows the substitution scheme of the ADAVR - approximate model, under the assumption that the losses in the magnetic core are not considered, the thyristors are assumed to be ideal switches, and the RC groups parallel to the thyristors do not exist. From the experiments and the computer simulations, n = 9 values are available for each type of relative absolute error, i.e. δ_{I1}, δ_{I2}, and δ_{U2} for the precise and the approximate models, or 6 error groups (Table 1).

Statistical processing of these data was performed assuming a finite number of values were available using the appropriate methodology [3, 4]. For this purpose, the following parameters were determined:

$$\bar{Y} = \frac{1}{n}.\sum_{k=1}^{n} Y_k , \qquad (2)$$

where $\bar{Y}$ is the arithmetic mean of the relative absolute error;

$$s(Y) = \sqrt{\frac{\sum_{k=1}^{n}(Y_k - \bar{Y})^2}{n-1}} , \qquad (3)$$

where s(Y) is the root mean square error;

$$\Delta Y = \frac{s(Y)}{\sqrt{n}}.t_p , \qquad (4)$$

where ΔY is the error of the arithmetic mean of the relative absolute error.

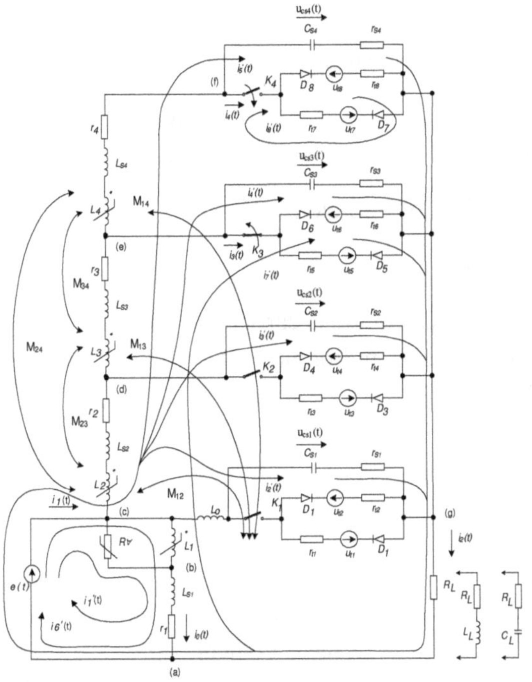

Fig. 1. Replacement scheme of ADAVR with four CTS (precise model).

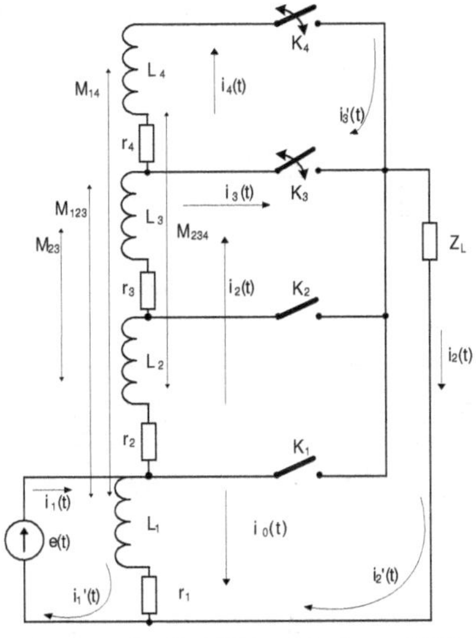

Fig. 2. Replacement scheme of ADAVR with four CTS (approximate model).

Here, t_p is a parameter of the Student's probability distribution, which is determined from appropriate statistical tables [3, 4], $\Delta_{1\sigma}(Y) = z_2.s(Y)$ is the maximum value of the of the root mean square error and $\Delta_{2\sigma}(Y) = z_1.s(Y)$ is the minimum value of the root mean square error.

Except that, $z_1 < z_2$, where z_1 and z_2 are coefficients from the relevant statistical tables, which are determined on the basis of the number of available values n, the coefficient of variation $w(Y) = \frac{S(Y)}{\bar{Y}}.100\ [\%]$ and the confidence probability P(Y) [3, 4].

As a result of the calculations carried out using the AVTOER1 program in the MATLAB integrated environment, the following data were obtained, presented in Table 2.

Table 1. Values of the relative absolute errors found as a result of the vector measurements.

	Precise model			***Approximate model***		
R_L	**δ_{I1}**	**δ_{I2}**	**δ_{U2}**	**δ_{I1}**	**δ_{I2}**	**δ_{U2}**
Ω	%	%	%	%	%	%
R load						
115,8	0,66	0,48	0,48	3,74	0,4	0,4
61,8	0,196	0,57	0,57	3,86	3,05	3,05
35,5	0,52	0,32	0,32	0,64	0,59	0,14
R-L load, L=1,0460H						
115,8	5,57	2,84	0,76	11,62	2,86	0,73
61,8	3,55	2,18	0,72	7,27	1,94	0,73
35,5	11,71	11,56	0,89	13,22	11,69	0,86
R-C load, C=38,09μF						
115,8	0,37	1,01	0,09	7,16	0,37	0,74
61,8	1,23	0,93	2,38	4,99	0,07	1,51
35,5	1,42	5,28	1,34	4,53	4,98	1,73

Table 2. Results of the relative absolute error analysis Y.

Quantity	*Precise model*			*Approximate model*		
Y [%]	δ_{I1}	δ_{I2}	δ_{U2}	δ_{I1}	δ_{I2}	δ_{U2}
$\bar{Y}$[%]	2,80	2,80	0,84	6,34	2,88	1,10
ΔY [%]	2,35	2,26	0,42	2,47	2,28	0,55
s(Y) [%]	3,78	3,65	0,68	3,99	3,68	0,88
$\Delta_{1\sigma}$(Y) [%]	6,47	6,24	1,16	6,82	6,30	1,15
$\Delta_{2\sigma}$(Y) [%]	2,72	2,62	0,49	2,86	2,64	0,63

The data presented in Table 2 should be considered as follows (for example for the first column of values, i.e. at $Y \equiv \delta_{I1}$):

$Y \equiv \delta_{I1} = (\bar{Y} \pm \Delta Y) = (\bar{\delta}_{I1} \pm \Delta\delta_{I1}) = (2{,}80 \pm 2{,}35)\ \%$, i.e. this is the interval estimate for the relative absolute error δ_{I1};

$\sigma(Y) = \sigma(\delta_{I1}) = [s(Y)^{+\Delta_{1\sigma}(\delta_{I1})}_{-\Delta_{2\sigma}(\delta_{I1})}] = (3{,}78^{+6{,}47}_{-2{,}72})\ \%$, i.e. this is the standard error (or so-called standard) of the probability distribution of the relative absolute error (in this case δ_{I1}).

An element of the computer simulation analysis is also the transient over-voltages $u_L(t)$ occurring during the switching of the CTS.

In the study of transients, the determination of initial conditions is possible under the following assumptions:

- $i_L(0-) = i_L(0+)$ and $u_C(0-) = u_C(0+)$, i.e. the so-called "correct set-up";
- $i_L(0-) \neq i_L(0+)$ and $u_C(0-) \neq u_C(0+)$, i.e. the so-called "incorrect set-up", in this $i_L(t)$ and $u_C(t)$ can change by jumps, and $i_C(t)$ and $u_L(t)$ at the moment of the step change turn out to be infinitely large.

The analysis of the commutation over-voltages in order to obtain a quantitative estimation relevant to the selection of the thyristors from the CTS and the sizing of the RC groups connected in parallel to the latter, was carried out by setting the initial conditions at "correct set-up" using the approximate model.

Fig. 3, 4 and 5 show the variation of $u_{L4}(t)$ for different phases of the switching process as the magnitude of the input supply voltage increases. Fig. 6, 7 and 8 show the variation of $u_{L4}(t)$ for different phases of the switching process as the magnitude of the input supply voltage decreases. In the first case L_4 receives a short circuit across two conducting thyristors from two adjacent CTSs and no over-voltages occur.

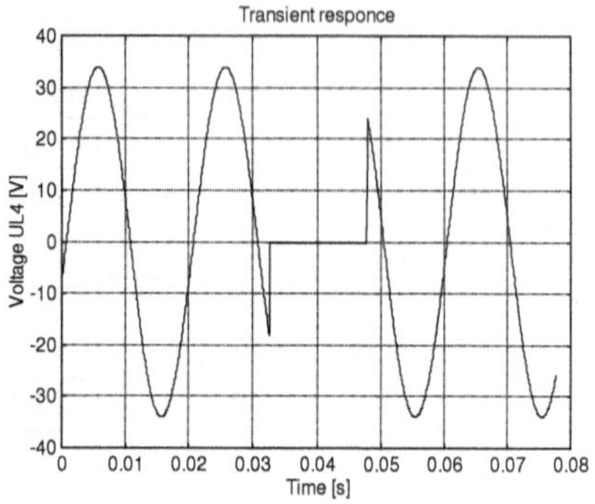

Fig. 3. Variation of $u_{L4}(t)$ with increasing input supply voltage from level 160V, angle of commutation $\varphi = 225^0$ and load $R_T = 35{,}55\Omega$.

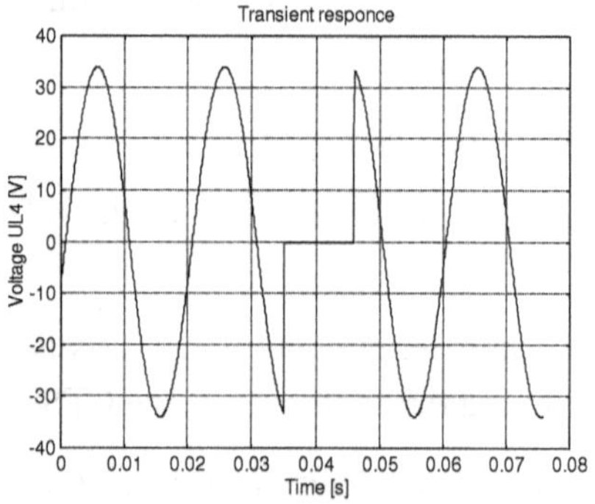

Fig. 4. Variation of $u_{L4}(t)$ with increasing input supply voltage from level 160V, angle of commutation $\varphi = 270^0$ and load $R_T = 35{,}55\Omega$.

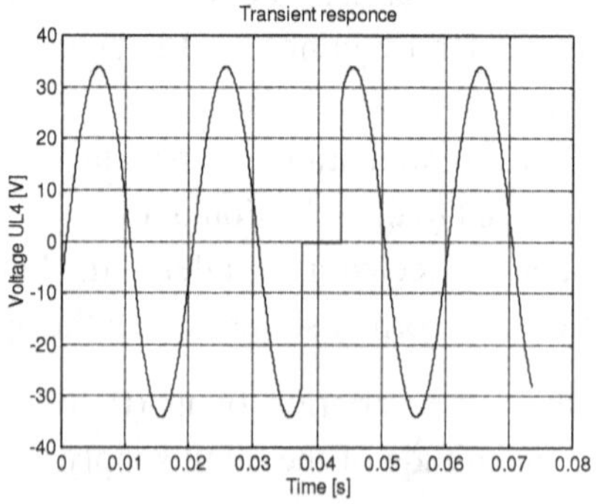

Fig. 5. Variation of $u_{L4}(t)$ with increasing input supply voltage from level 160V, angle of commutation $\varphi = 315^0$ and load $R_T = 35{,}55\Omega$.

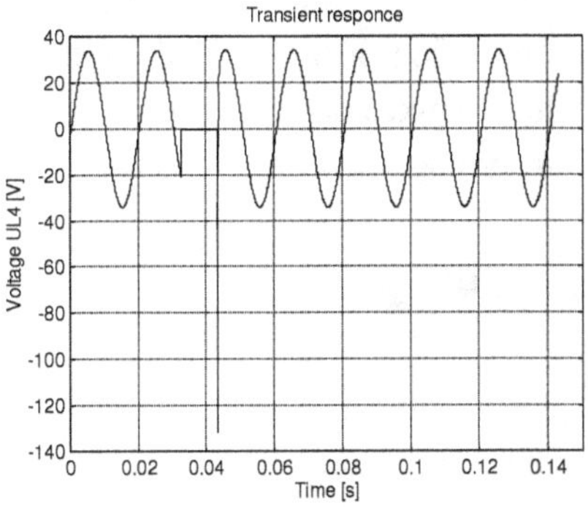

Fig. 6. Variation of $u_{L4}(t)$ with decreasing input supply voltage from level 160V, angle of commutation $\varphi = 225^0$ and load $R_T = 35,55\Omega$.

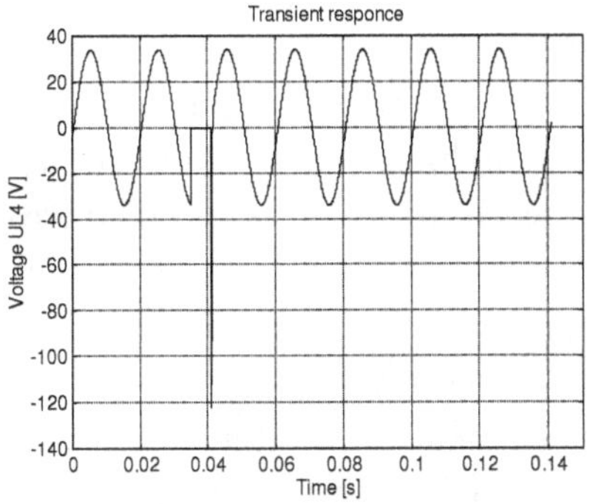

Fig. 7. Variation of $u_{L4}(t)$ with decreasing input supply voltage from level 160V, angle of commutation $\varphi = 270^0$ and load $R_T = 35,55\Omega$.

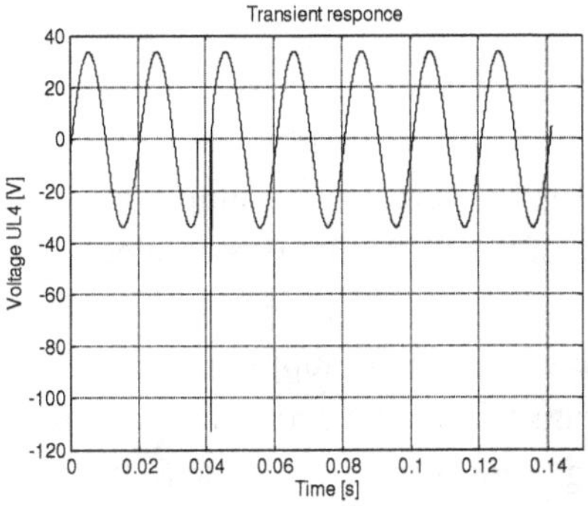

Fig. 8. Variation of $u_{L4}(t)$ with decreasing input supply voltage from level 160V, angle of commutation $\varphi = 315^0$ and load $R_T = 35,55\Omega$.

In the second case – when the magnitude of the input supply voltage decreases when the thyristor of the CTS to be switched on is released, conditions are created for forced commutation of the conducting thyristor of the CTS to be switched off. The overvoltage spikes occurring as a function of the phase of the start of the commutation process are in the specific cases with amplitude 110V÷130V.

III. CONCLUSIONS

Comparing the results of Table 2 for the precise model and the approximate model leads to the following conclusions:

1. The arithmetic mean errors and standard errors for the precise model are smaller than those for the approximate model, i.e., the precise model gives higher confidence in the results of computer simulations compared to the approximate model. However, this comes at the expense of increased machine computation time and occupation of greater computer system resources. At the same time, the approximate model leads to larger errors, but the result is obtained faster.
2. In the precise model, the arithmetic mean error does not exceed 2,80% and in the approximate model it reaches 6,34%. The standard error of the precise model does not exceed 3,78%, while that of the approximate model is 3,99%.
3. The precise model is more sophisticated than the approximate model because it takes into account a larger number of ADAVR parameters and allows simulation of a larger number of quantities (for example, the currents and voltages in the switch-off RC groups to the switching elements, something that cannot be tracked with the approximate model).
4. The error analysis performed to simulate the processes in the ADAVR allows to make a model selection with a view to design a specific regulator for a specific application problem.
5. From the simulations, it is clear that the voltage $u_{L4}(t)$ contains an impulse (switching overvoltage) during switching as a result of a decrease in the magnitude of the input supply voltage. The amplitude of this pulse is dependent on the phase of the onset of the switching process and in the case studies is in the range 110V÷ 130V.
6. When the magnitude of the input supply voltage increases, $u_{L4}(t)$ does not contain a commutation pulse, but there occurs a short circuit of the corresponding autotransformer winding for a time up to 10mS [5], which implies a current overload of the power circuit elements relative to the established mode.

REFERENCES:

[1]. Barudov E., Panov E., Barudov S., Analysis of Electrical Processes in Alternating Voltage Control Systems. Journal of International Scientific Publication: Materials, Methods & Technologies, 2010, Vol. 4, Part 1, pp. 154÷182, ISSN 1313-2539.

[2]. Barudov E., Panov E., Barudov S., Comparative study of the switching processes in autotransformer discrete step-up voltage regulator. ELMA 2011, pp. 55÷60, ISSN 1313-4965.

[3]. Rumshisky L. Z., Mathematical processing of experimental results, Nauka, Moscow, 1971.

[4]. Armutliysky D., A short guide to mathematical processing of experimental data, VMEI Varna, 1974.

[5]. Barudov S., Panov E., Barudov E., Study of the sensitivity of the models of autotransformer discrete alternating voltage regulator, Marine Scientific Forum, Electronics, Electrical Engineering and Automatics. Informatics, VVMU "N. Y. Vaptsarov", Vol. 4, 2013, pp. 97-103, ISSN 1310-9278.

Contact:

Assoc. Prof. PhD. Eng. Emil Panov, Bulgaria, Varna 9010, Technical University, 1 Studentska str., Department TIE, e-mail: eipanov@yahoo.com.

Chief Ass. Prof. Emil Barudov, Bulgaria, Varna 9010, Technical University, 1 Studentska str., Department TIE, e-mail: ugl@abv.bg.

Prof. PhD Eng. Stefan Barudov, Bulgaria, Varna 9026, VVMU "N. Y. Vaptsarov", 73 Vasil Drumev str., Department ET, e-mail: sbarudov@abv.bg.

Chapter 7. Results of the Experiments of ADAVR with R, RL and RC Loads

Exploration of the Electric Processes in Discrete Alternating Step-Voltage Regulators

Emil PANOV
Department of Theoretical Electrical Engineering and Instrumentation
Technical University of Varna
9010, Varna, Bulgaria
e-mail: eipanov@yahoo.com
Emil BARUDOV
Department of Electrical Engineering
Nikola Y. Vaptsarov Naval Academy
9000, Varna, Bulgaria
e-mail: ugl@abv.bg
Milena IVANOVA
Department of Electric Power Engineering
Technical University of Varna
9010, Varna, Bulgaria
e-mail: m.dicheva@tu-varna.bg
(Published in the Proceedings of the XX-th International Symposium on Electrical Apparatus and Technologies SIELA'2018, Bourgas, Bulgaria, 3-6 June 2018, pp. 325 - 328, ISBN 978-1-5386-3418-9)

Abstract—**The paper presents a review on the problems of the analysis of autotransformer discrete alternating voltage regulators. The analysis has been made basically by the state variables approach in the time domain, where the non-linearity of the regulator components has been considered. A special algorithm has been developed for part-by-part solving of the obtained sparse matrix equations with cell structure. Experimental verification with a prototype has been performed, which showed a good correspondence with the simulation results.**

Keywords—commutation process; discrete alternating voltage regulator; semiconductor commutation elements; transient processes.

I. Introduction

An important requirement for the electric power quality is the range of alteration of the input supply voltage amplitude. One of the ways for meeting this requirement is by implementation of autotransformer discrete alternating voltage regulators (ADAVR) with semiconductor commutation elements (SCE) [1, 2]. The design and the construction of such regulators require modern approaches for computer analysis and simulation of the complex processes occurring in these devices [3, 4]. The major problem is the development of an adequate model and algorithms for analysis, which would enable the achievement of accurate solutions and true simulations of the ADAVR responses in a variety of operative modes and in emergent situations as well [5, 6]. The final goal of the whole research is to develop and to implement a procedure for virtual and practical design of ADAVR.

II. Analysis

In Fig. 1 an equivalent circuit of ADAVR with four SCE is shown. The circuit includes the parameters of the magnetic circuit, the commutation groups, the switch-off assemblies, the parameters of the separate sections of the winding and it takes into account the existing non-linearities. The created model enables detection of random commutations of adjacent SCE in the time domain.

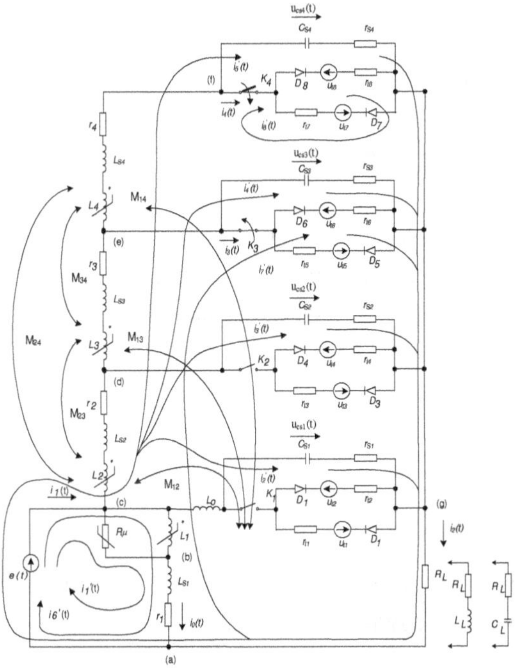

Fig. 1. Equivalent circuit of ADAVR with four SCE.

The commutation process includes the commutation of two adjacent switches, and the increasing or the decreasing input supply voltage is explored in three time intervals. In the first interval ADAVR operates in stationary AC regime as the electric equilibrium is described by the loop currents method in phasor form. The general system of the equations assumes the following phasor matrix form:

$$[Z_l'].[\dot{I}'] = [\dot{E}], \tag{1}$$

where $[Z_l']$ is the contour impedance matrix of ADAVR with a dimension of 7 rows by 7 columns (i.e. 7x7); $[\dot{I}']$ is a matrix-column of the loop currents $\dot{I}_1', \dot{I}_2', ..., \dot{I}_7'$ with a dimension of 7x1; $[\dot{E}]$ is a matrix-column of the contour emf with a dimension 7x1. The numerical method, chosen for solving (1), is the method of Gauss-Jordan. The matrix $[Z_l']$ is not singular, irrespective of the variant of selection of the independent loops of the circuit in Fig. 1.

During the second interval, the first transient process in ADAVR occurs provided the switches K_3 and K_4 are closed. The analysis is conducted by the state variables approach. The general matrix form of the system of equations is as follows:

$$[A_2]\frac{dx}{dt}[x(t)] = [B_2].[x(t)] + [e_2(t)], \tag{2}$$

where the matrices $[A_2]$ and $[B_2]$ have dimensions 12x12 and they consist of nine submatrices.

During the third interval, the second transient process occurs, but here only the switch K_4 is closed. The analysis is conducted again by the state variables approach in the time domain. The system of equations in the third interval has the following form:

$$[A_3]\frac{dx}{dt}[x(t)] = [B_3].[x(t)] + [e_3(t)], \tag{3}$$

where the matrices $[A_3]$ and $[B_3]$ have dimensions 11x11 and each of them consists of three submatrices.

The stability of the solutions obtained by this algorithm does not depend on the commutation phase of switches K_3 and K_4.

An automated computer program AVTO was developed for simulation of the processes in ADAVR with SCE in the environment of MATLAB. The computer program comprises the specifics of the circuit design, the alteration of the parameters of the autotransformer as a function of the mode, the type and the parameters of the load, the initial phase of commutation and the parameters of the used semiconductor switches.

The practical realization of a discrete voltage regulator with two commutating SCE (the number of the switches does not change the scheme operating algorithm) is shown in Fig. 2.

A control algorithm for avoiding the simultaneous switching on of two opposed connected to the load thyristors from two neighbouring SCE is implemented for the commutations of the switches.

The results from the computer simulations and the experimentally recorded ones for the case of active load and commutation angle $\varphi=270^0$ are shown in Fig. 3a and Fig. 3b.

The commutation angle at different loads is chosen randomly and indicates the model adequacy. The experimental researches and the computer simulations for the currents flowing through the switches K_3 and K_4 for RL and RC loads are shown in Fig. 4a, Fig.4b and Fig. 5a, Fig. 5b, respectively. The obtained experimental values for the currents and the output voltage in stationary regime (closed switch K_3) are presented in Table 1.

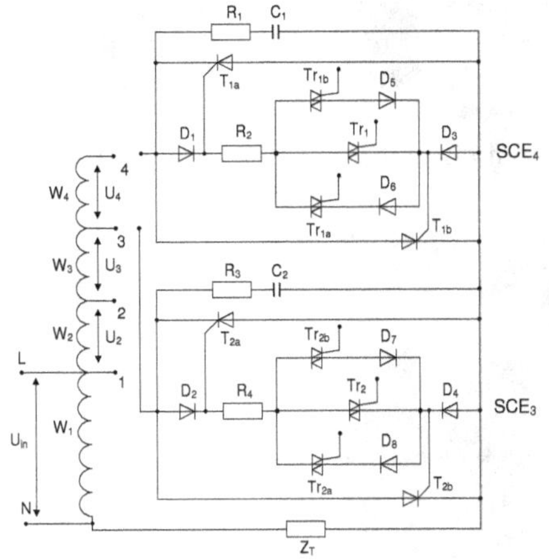

Fig. 2. Circuit diagram of ADAVR with two SCE.

TABLE 1.

Load	*Experimental data*			*Computer simulations*		
	I_1	I_2	U_2	I_1	I_2	U_2
	A	*A*	*V*	*A*	*A*	*V*
R	8,45	6,16	219	8,3965	6,1205	218,6
RL	9,75	7,03	219,9	8,5181	6,2016	217,8
RC	2,85	2,15	220,3	2,8034	2,1421	222,6

• Note*:* The used loads for the experiments and the simulations, shown in the table, are as follows:
R load – R = 35,5Ω;
RL load – R = 3,5Ω, L = 1,76H (connected in parallel);
RC load – R = 33,5Ω, C = 38,09μF (connected in series).

The numeric values received from the simulations by the used model of ADAVR does not differ with more than 4% of the experimental values given in the table.

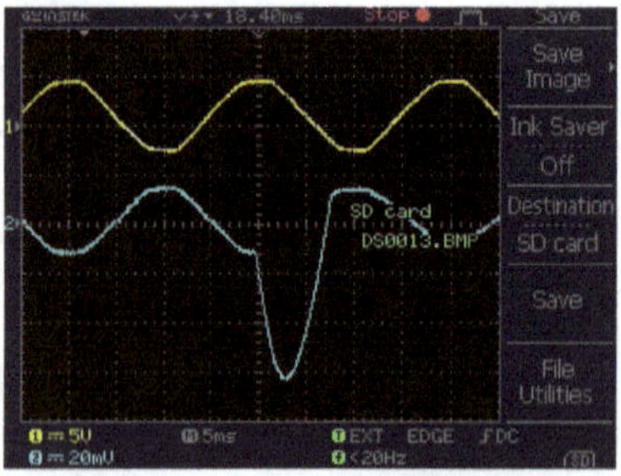

Oscillogram of the input current $i_1(t)$.

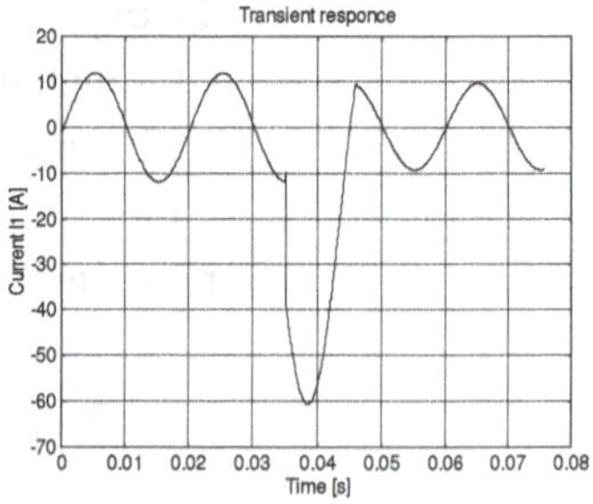

Simulation of the input current $i_1(t)$.

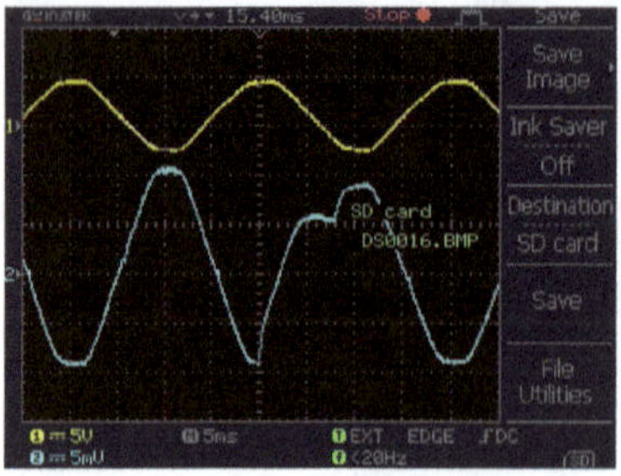

Oscillogram of the output current $i_2(t)$.

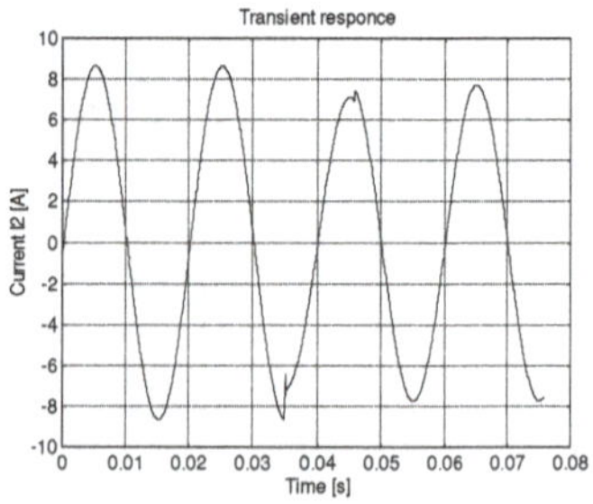

Simulation of the output current $i_2(t)$.

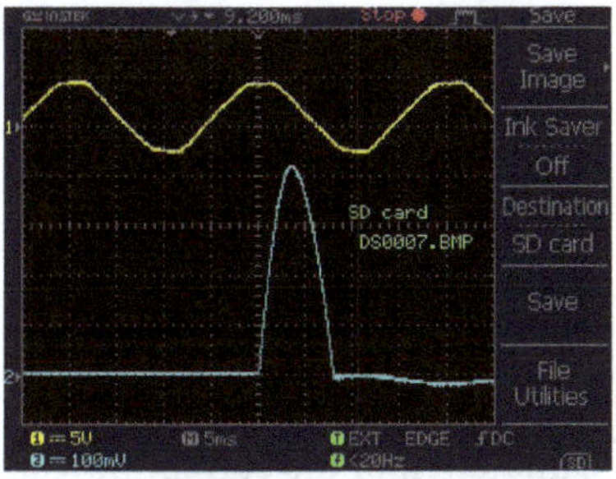

Oscillogram of the current flowing through switch K_3 - $i_3(t)$.

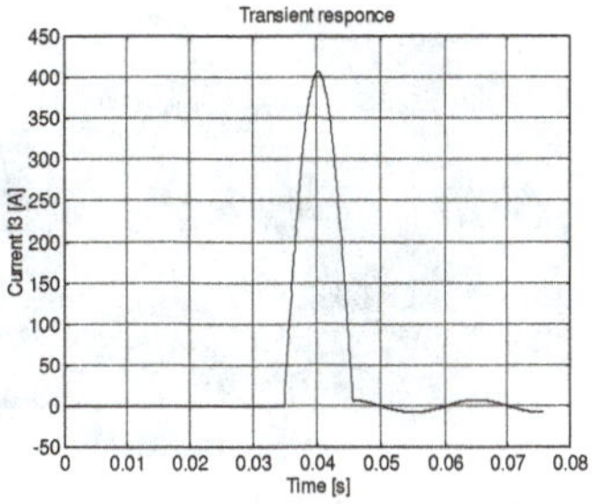

Simulation of the current flowing through switch K_3 - $i_3(t)$.

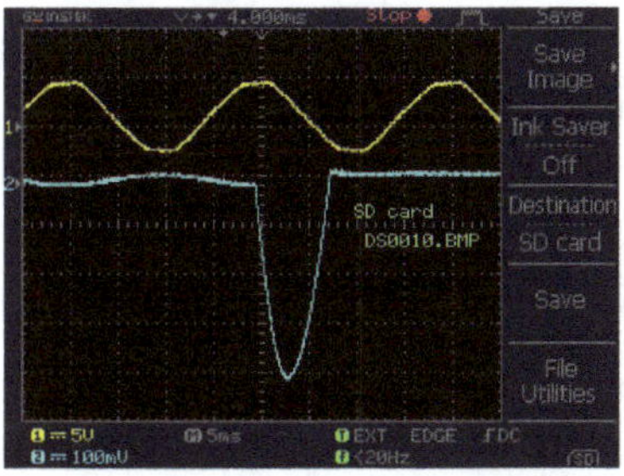

Oscillogram of the current flowing through switch K_4 - $i_4(t)$.

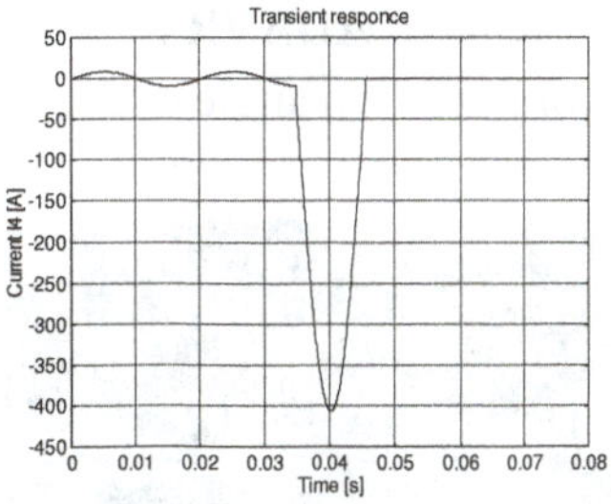

Simulation of the current flowing through switch K_4 - $i_4(t)$.

Fig. 3a. Experimental oscillograms of the currents of ADAVR for R load and commutation angle $\varphi=270^0$.

Fig. 3b. Computer simulations of the currents of ADAVR for R load and commutation angle $\varphi=270^0$.

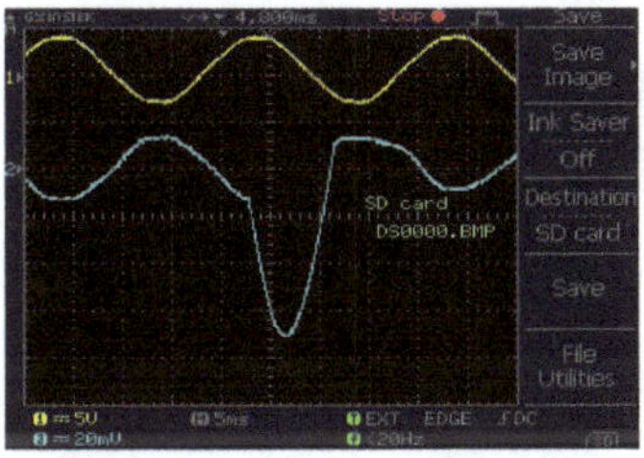

Oscillogram of the input current $i_1(t)$.

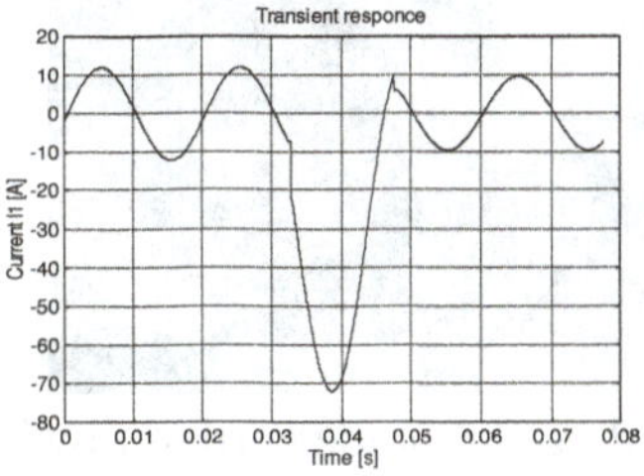

Simulation of the input current $i_1(t)$.

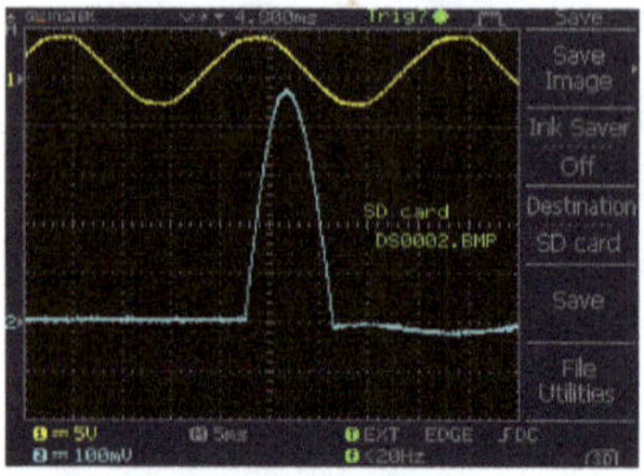

Oscillogram of the current flowing through switch K_3 - $i_3(t)$.

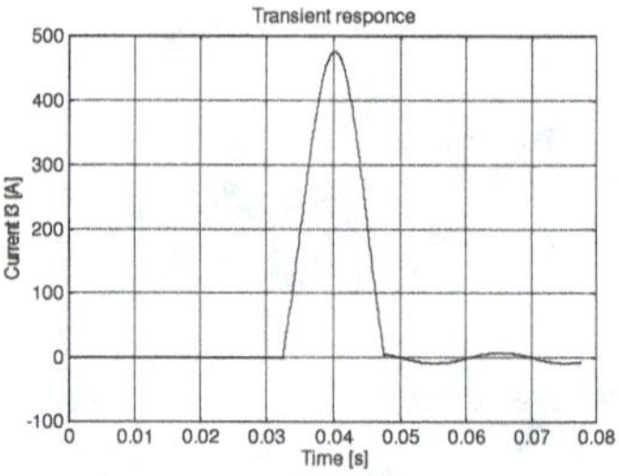

Simulation of the current flowing through switch K_3 - $i_3(t)$.

Fig. 4a. Experimental oscillograms of the currents of ADAVR for RL load and commutation angle φ=225.

Fig. 4b. Computer simulations of the currents of ADAVR for RL load and commutation angle $\varphi=225^0$.

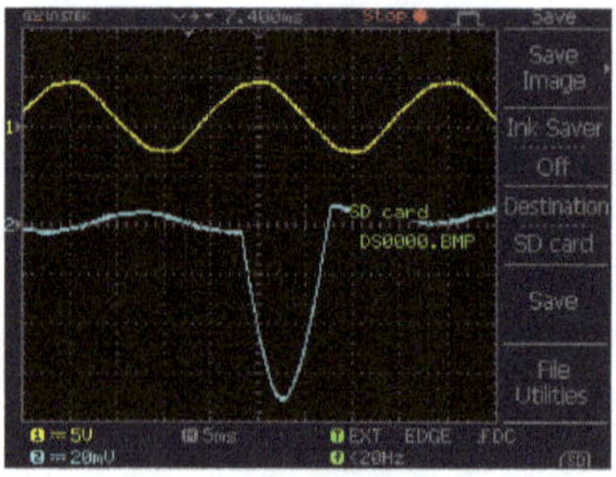

Oscillogram of the input current $i_1(t)$.

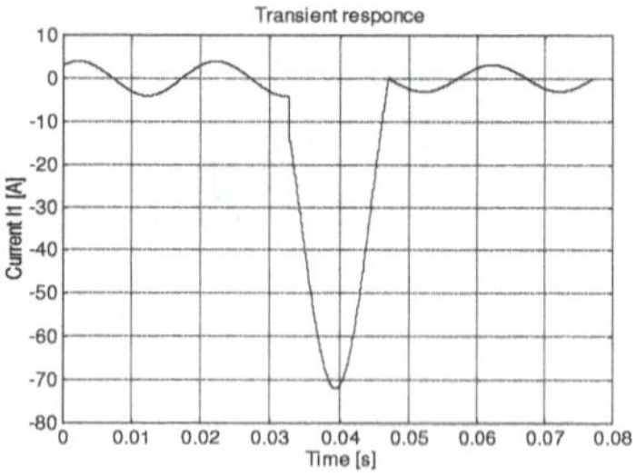

Simulation of the input current $i_1(t)$.

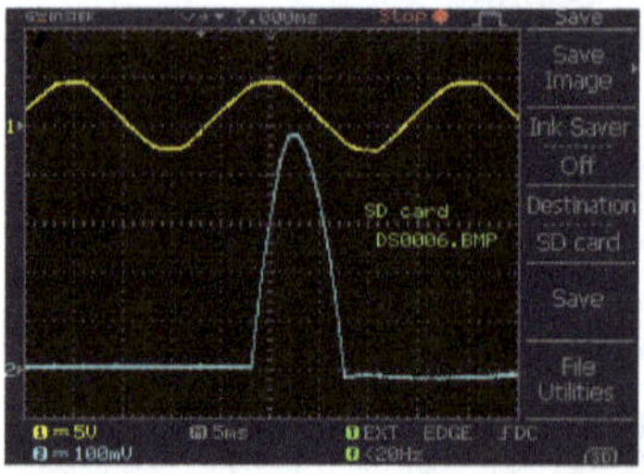

Oscillogram of the current flowing through switch K_3 - $i_3(t)$.

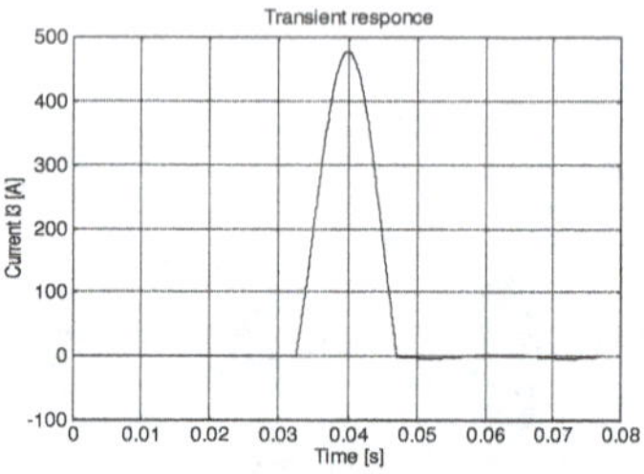

Simulation of the current flowing through switch K_3 - $i_3(t)$.

Fig. 5a. Experimental oscillograms of the currents of ADAVR for RC load and commutation angle φ=225.

Fig. 5b. Computer simulations of the currents of ADAVR for RC load and commutation angle $\varphi=225^0$.

III. **Conclusions**

1. A mathematical model for description of the commutation processes in ADAVR and a detailed algorithm for solving the system of equations, describing the electrical equilibrium in the voltage regulator, are proposed.
2. A computer program AVTO has been developed for simulation of the complex processes occurring in ADAVR during the commutation processes, which enables a complete analysis of the explored devices.
3. The conducted analytical and experimental researches with different phases of commutation and types of the load show a very good correspondence of the obtained results, which demonstrates the adequacy of the model and gives the opportunity of virtual design of ADAVR.

References

[1] S. Barudov, and E. Barudov, Discrete Alternating Current Regulators and Stabilizers, Sofia-Moscow, Pensoft, 2006, ISBN 978-954-642-276-7.

[2] E. Barudov, Exploration and Analysis of the Electrical Processes in Circuits and Devices for Discrete Regulation of the Magnitude of AC Voltage, Auto-essay on PhD dissertation, "Nikola Vaptsarov" Naval Academy, Varna, 2014. (in Bulgarian)

[3] S. Barudov, E. Panov, E. Barudov, and M. Ivanova, Discrete stabilizer of AC voltage, Patent BG1727 U1, 2013. (in Bulgarian)

[4] E. Barudov, E. Panov, and S. Barudov, Commutation processes with reactive loads in discrete alternating voltage regulator, Annual of the Technical University of Varna, 2011, pp. 3-8, ISSN 1311-896X. (in Bulgarian)

[5] E. Panov, E. Barudov, and S. Barudov, Vector analysis and comparative valuation of precise and approximate non-linear models of discrete regulator with reducing input AC voltage, XLVIII International Scientific Conference on Information, Communication and Energy Systems and Technologies ICEST'2013, Macedonia, Vol. 2, 2013, pp. 771-774, ISBN 978-9989-786-89-1.

Chapter 8. Exploration on the Efficiency of the ADAVR

Study of the Electrical Characteristics of Autotransformer Discrete Alternating Voltage Regulators with R-L Loads

Emil PANOV*, Milena IVANOVA* and Emil BARUDOV**

*Technical University of Varna, Faculty of Electrical Engineering, 9010 Varna, Bulgaria, e-mail: eipanov@yahoo.com
e-mail: m.dicheva@tu-varna.bg
** Nikola Vaptsarov Naval Academy, Department of Electrical Engineering, 9000 Varna, Bulgaria,
e-mail: ugl@abv.bg
(Published in the Proceedings of the XVI-th International Conference on Electrical Machines, Drives and Power Systems ELMA'2019, 6-8 June 2019, Varna, Bulgaria, pp. 321-325, ISBN 978-1-7281-1412-5.)

Abstract—**Autotransformer discrete alternating voltage regulators (ADAVR) are used for limitation of the supply voltage range and thus prevent the deterioration of the electrical energy quality. The paper is dedicated to the investigation of the processes in an ADAVR with R-L load. The study was conducted by the help of an automated computer program in the environment of MATLAB, allowing simulation of the processes in ADAVR at different switching ranges corresponding to different supply voltages. The results obtained for the regulator parameters (efficiency, input and output currents, output voltage, input and output active and reactive power) were verified experimentally with a prototype of ADAVR and a good correspondence with the data from the simulated processes was obtained.**

Keywords— ***autotransformer discrete alternating voltage regulator, efficiency, semiconductor commutation elements, thyristor, commutation process.***

I. Introduction

In the last years there has been a strong interest in more accurate control of the parameters of the supplied electrical energy for industrial and household use. An effective approach for improving the electrical energy quality is by regulating the amplitude of the input supply voltage, which is often preferred by the use of an ADAVR with semiconductor commutation elements (SCE) [1, 2, 3, 4, 5]. In previous researches ADAVR with R-loads had been examined [6, 7]. In some practical application such regulators are loaded sometimes with R-L or R-C loads depending on the type of the connected consumers. This requires conducting of computer simulations of the electrical processes in ADAVR on the basis of developed mathematical models [2, 8, 9, 10, 11, 12, 13].

For the present study, the regulator was loaded by a load resistance R_L, inductive load L and R_{st}, which considers the losses in the steel core of the coil - all connected in parallel (Fig. 1). The replacement scheme of one SCE with thyristors is shown in Fig. 2. For the analysis of ADAVR the parallel R-L load is transformed into a serial scheme. An experimental study of the processes in ADAVR with R-L load was performed for verification of the mathematical model authenticity and the simulation results.

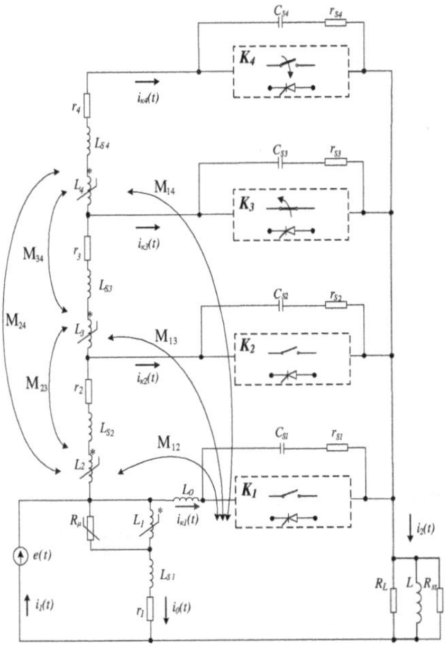

Fig. 1. Equivalent circuit of ADAVR with four semiconductor commutation elements K_1-K_4.

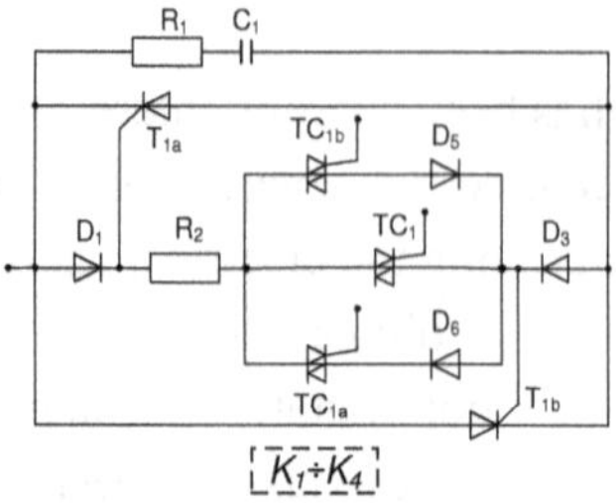

Fig. 2. Replacement scheme of a single SCE.

The computer simulations for the ADAVR with R-L load were performed, based on the analysis of the electrical processes in the equivalent circuit of ADAVR with four SCE (Fig. 1), including the parameters of the ferromagnetic core of the autotransformer, the commutation groups, the switch-off assemblies, the parameters of the separate sections of the winding. The existing non-linearities were also considered. The specifics of the commutation of consecutively switched neighboring thyristor switches had been taken into account for the creation of the mathematical model as the commutation process included three separate time intervals [7]:

- First interval: ADAVR operates in stationary AC regime with one closed SCE as the electric equilibrium is described by the help of the loop analysis in phasor form.
- Second interval: the first transient process in ADAVR occurs, provided two of the SCE switches are closed. The analysis is conducted by the state variables approach in the time domain.
- Third interval: the second transient process occurs, but only one SCE is closed. The analysis is conducted again by the state variables approach in the time domain.

The detailed mathematical description of the electrical processes in the three intervals was presented in [2, 5].

II. Simulation results

Simulations of the processes in ADAVR with R-L load (Fig. 1) in stationary AC regime for four operating regimes, (when only one of the four switches K_1, K_2, K_3 or K_4 is closed) were performed. The parameters, set in the simulation program [2], are: $R_L = 3{,}4921\Omega$, $5{,}5\Omega$, 11Ω, 22Ω, $3\ 5{,}552\Omega$; $L = 0{,}05H$; $R_{steel} = 3438{,}3\Omega$. The alterations of the input current I_1, the output voltage U_2, the output current I_2, the input real power P_{gen}, the input reactive power Q_{gen}, the output real power P_2, the output reactive power Q_L and the efficiency η as functions of the load resistance R_L were obtained.

The input voltage E for the four studied cases was as follows:

- K_1 – closed – $E = 220 \div 200{,}1$V;
- K_2 – closed – $E = 200 \div 180{,}1$V;
- K_3 – closed – $E = 180 \div 160{,}1$V;
- K_4 – closed – $E = 160 \div 150$V.

The graphical results from the simulations for the case, when switch K_3 is closed, are presented in Fig. 3÷9, where:

- when E = 180V – symbol "star" was used;
- when E = 160,1V – symbol "circle" was used.

The numerical results for the studied electrical characteristics for all cases are shown in Table 1.

It is well seen, that Q_{gen} drops down with the decrease of the value of the input voltage E within the range of each of the cases (Table 1). The input real power P_{gen}, the input reactive power Q_{gen} and the output real power P_2 decrease with the increase of the value of the load resistance R_L for each of the cases. The reactive power of the load Q_L slightly increases with the increase of R_L at a constant value of E. U_2 is also almost constant in the range of changing of the load resistance R_L for the first two cases.

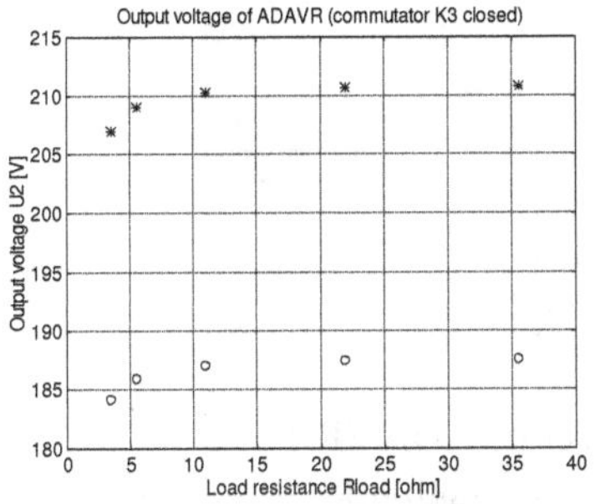

Fig. 3. Output voltage of ADAVR U_2 versus the load resistance R_L when switch K_3 is closed.

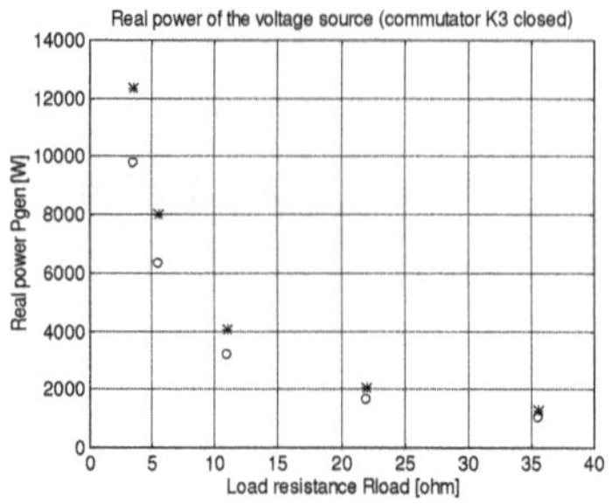

Fig. 4. Input real power of ADAVR P_{gen} versus the load resistance R_L when switch K_3 is closed.

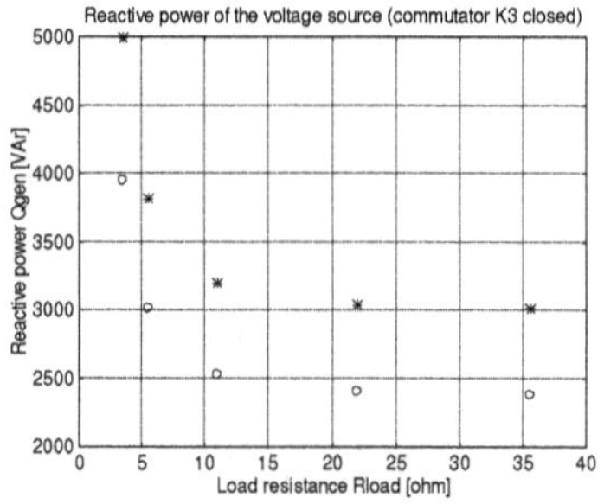

Fig. 5. Input reactive power of ADAVR Q_{gen} versus the load resistance R_L when switch K_3 is closed.

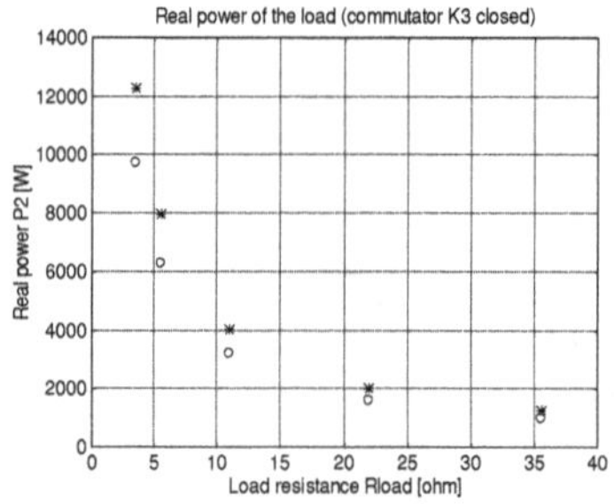

Fig. 6. Output real power of ADAVR P_2 versus the load resistance R_L when switch K_3 is closed.

TABLE 1. SIMULATION RESULTS OF ADAVR FOR THE DIFFERENT CASES OF CLOSED SCE

Closed SCE	K_1		K_2		K_3		K_4	
Input voltage E, V	220	200,1	200	180,1	180	160,1	160	150
Input real power P_{gen}, kW	14÷1,4	11,5÷1,2	13,6÷1,4	11,1÷1,1	12,4÷1,28	9,79÷1,01	7,56÷1,035	6,64÷0,9
Input reactive power Q_{gen}, VAr	3291,3	2722,8	3526÷3172	2860÷2572	4989÷3011	3947÷2382	7144÷2728	6279÷2398
Input current I_1, A	65÷16	59÷15	70,5÷17,3	63,46÷15,57	74,1÷18,18	65,9÷16,17	65÷18,24	60,94÷17,1
Output real P_2, kW	14÷1,4	11,5÷1,1	13,6÷1,35	11,02÷1,1	12,3÷1,26	9,71÷0,999	7,45÷1,017	6,55÷0,893
Output reactive power Q_L, VAr	3081,2	2549	3019,4÷3033,9	2448,5÷2460,2	2726÷2828	2157÷2237	1655÷2278	1455÷2002
Output voltage U, V	220	200,1	217,78÷218,3	196,11÷196,58	207÷211	184÷187,46	161÷189,16	151,2÷177
Output current I_2, A	64÷15	59÷14	63,95÷15,22	57,59÷13,7	60,77÷14,7	54,05÷13,1	47,35÷13,2	44,4÷12,36
Efficiency η, %	99,831÷98,317		99,638 (max)÷98,598		99,377 (max)÷98,47		99,1 (max) ÷98,256	

The efficiency of the ADAVR η was calculated according to (1). It depends only on the parameters of the regulator and the resistance of the load R_L, i.e. η depends on the real powers at the input and at the output of the regulator:

$$\eta = \frac{P_2}{P_{gen}}.100\% \,. \tag{1}$$

The results show that η is decreasing with the increase of R_L, when switch K_1 is closed. The efficiency η has a maximum with the increase of the value of the load resistance R_L, when only one of the switches K_2, K_3 or K_4 is closed (Fig. 9). The dependency of the efficiency η on the values of the load resistance R_L is similar for all values of the input voltage E, corresponding to different SCE closed.

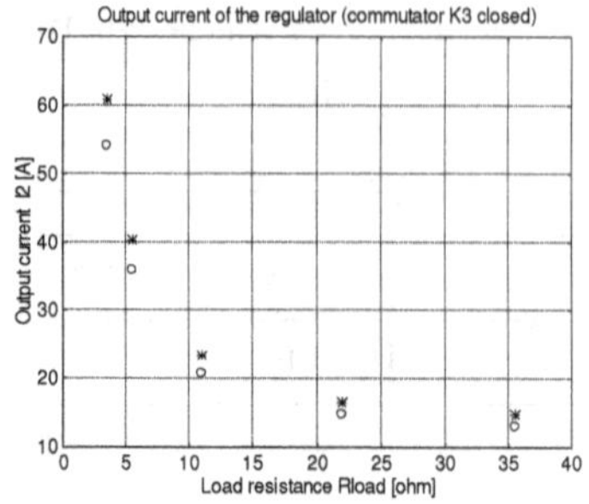

Fig. 7. Output current of ADAVR I_2 versus the load resistance R_L when switch K_3 is closed.

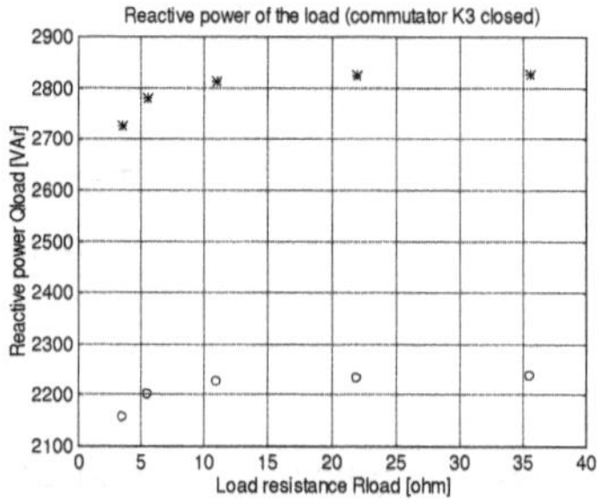

Fig. 8. Output reactive power of ADAVR Q_L versus the load resistance R_L when switch K_3 is closed.

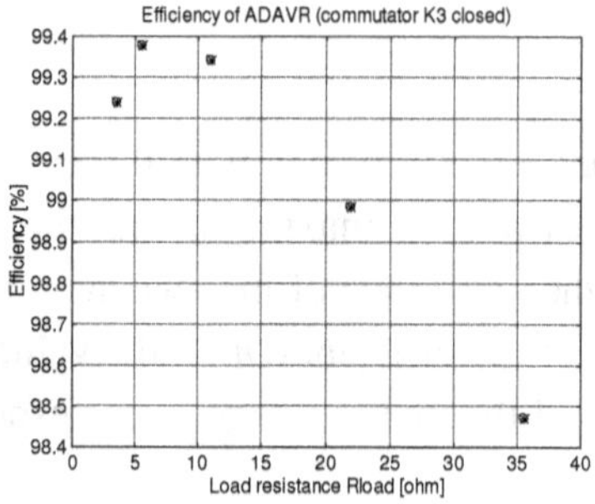

Fig. 9. Efficiency coefficient of ADAVR η versus the load resistance R_L when switch K_3 is closed.

The efficiency varies from 98,256% to 99,831% for the different cases (Table 1).

III. Experimental verification

The results from the simulations have been experimentally verified and showed a very good correspondence. The comparison of the results from the physical experiments and the computer simulations [14] in Table 2 indicate that the relative error is less than 1,32% for the measured quantities. The oscillograms of the input current $i_1(t)$ and the output current $i_2(t)$ at R_L=61,8Ω, L=2,1712H, when the input voltage E drops down from 160,1V to 159V, are presented in Fig. 10a and Fig. 11a [15]. The corresponding results from the computer simulations are presented in Fig. 10b and Fig. 11b [15].

TABLE 2. COMPARISON OF THE RESULTS FROM THE EXPERIMENTS AND THE SIMULATIONS

	Experimental results				***Computer simulations***			
R_T,	U_1	I_1	U_2	I_2	U_1	I_1	U_2	I_2
Ω	V	A	V	A	V	A	V	A
61,8	160	5,1	220,1	3,71	160	5,167	217,46	3,6921

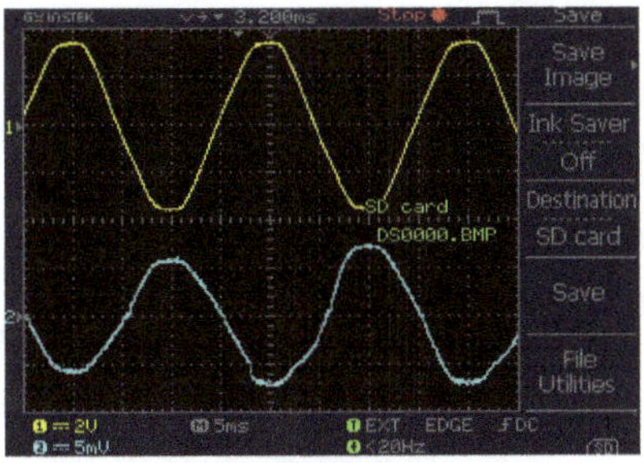

Fig. 10a. Oscillogram of the input current $i_1(t)$ at R_L=61,8Ω, L=2,1712H, E=160V.

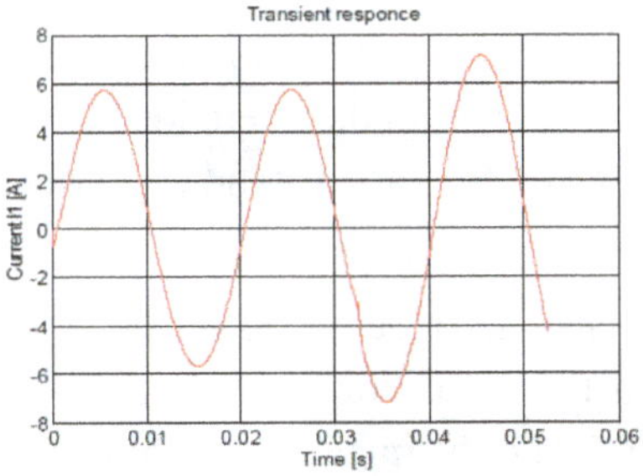

Fig. 10b. Simulation results for the input current $i_1(t)$ at R_L=61,8Ω, L=2,1712H, E=160V.

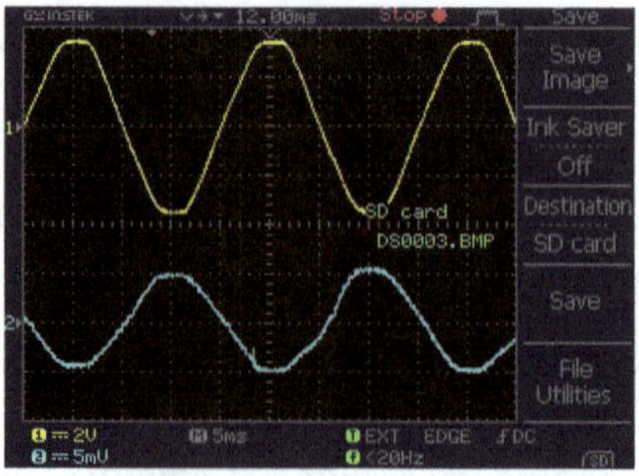

Fig. 11a. Oscillogram of the output current $i_2(t)$ at R_L=61,8Ω, L=2,1712H, E=160V.

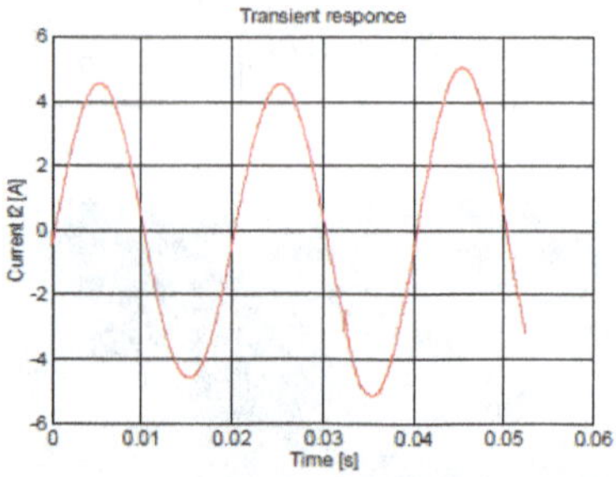

Fig.11b. Simulation results for the output current $i_1(t)$ at R_L=61,8Ω, L=2,1712H, E=160V.

IV. Conclusions

1. The efficiency of the ADAVR η varies from 99,831% to 98,256%, which shows a very good performance of the explored prototype.

2. Regarding the quality of electrical power at the regulator's output, the change of the output voltage U_2 with the alteration of the load resistance R_L fully meets the requirements of the Bulgarian standard, when the input voltage is not less than E = 170V.

3. Regarding the consumed reactive power Q_{gen} at the input of the regulator, it is obvious, that when the input voltage E is less than 180,1V and the load resistance R_L is less than 5,5Ω, the input reactive power becomes a factor, which must be taken seriously into account.

References

[1] S. Barudov, E. Panov, and E. Barudov, "Parameters of the commutation process in a step-voltage AC regulator," Proceedings of Sankt Petersburg Institute for Qualification Promotion "Methods and Tools for Estimation of the State of Power Equipment", Sankt Petersburg, Russia, vol. 25, pp. 67-76, 2005. UDK 621.3.048, 621.315.62, BBK 31.264-04 (in Russian)

[2] E. Barudov, E. Panov, and S. Barudov, "Analysis of electrical processes in alternating voltage control systemsm," Journal of International Scientific Publications: Materials, Methods & Technologies, Bulgaria, vol. 4, part 1, pp. 154 – 182, 2010. ISSN: 1313-2539.

[3] E. Panov, E. Barudov, and S. Barudov, "Vector analysis and comparative valuation of precise and approximate non-linear models of discrete regulator with reducing input AC voltage," Proceedings of XLVIII-th International Scientific Conference on Information, Communication and Energy Systems and Technologies ICEST'2013, Ohrid, Macedonia, vol. 2, pp. 771-774, 2013. ISBN 978-9989-786-89-1.

[4] S. Barudov, E. Panov, E. Barudov, and M. Ivanova. Discrete stabilizer of AC voltage, Patent BG1727 U1 2013. (in Bulgarian)

[5] E. Barudov, Exploration and Analysis of the Electrical Processes in Circuits and Devices for Discrete Regulation on the Magnitude of AC Voltage, Auto-essay on PhD dissertation, "Nikola Vaptsarov" Naval Academy, Varna, 2014. (in Bulgarian)

[6] E. Panov, E. Barudov, and M. Ivanova, "Exploration of the electric processes in discrete alternating step-voltage regulators," Proceedings of the XX-th International Symposium on Electrical Apparatus and Technologies SIELA, Bourgas, Bulgaria, pp. 325-328, 2018. ISBN 978-1-5386-3418-9.

[7] M. Ivanova, "Exploration of the efficiency of autotransformer discrete alternating voltage regulators," Proceedings of the Union of Scientists – Varna, series "Technical Sciences", vol. 1, pp. 21-29, 2018. ISSN 1310-5833.

[8] N. Djagarov, Zh. Grozdev, and M. Bonev, "Use of adaptive system stabilizers in ship power systems," Proceedings of 15th IEEE International Conference on Environment and Electrical Engineering, Rome, Italy, pp. 593-398, 2015.

[9] N. Djagarov, Zh. Grozdev, and M. Bonev, "Mathematical modeling of electromechanical processes in ship's power systems," Proceedings of 15-th IEEE International Conference on Environment and Electrical Engineering, Rome, Italy, pp. 1155-1162, 2015.

[10] M. Mehmed-Hamza and N. Nikolaev, "Using Simulink models for education purposes in the subject automation of electric power systems," Proceedings of 18-th International Symposium on Electrical Apparatus and Technologies, SIELA'2014, pp. 137–140, 2014.

[11] M. Hamza, M. Vasileva, and M. Yordanova, "Co-ordination of the operation of the relay protection and surge protective devices in electrical power networks medium voltage 20 kV," *J. Electr. Eng.*, vol. 60, № 3, pp. 170–172, 2009.

[12] M. Mehmed-Hamza, M. Vasileva, and P. Stanchev, "Increasing the education quality by means of computer-aided visualization of the processes in electric power systems,“ Proceedings of the Second International Scientific Conference "Intelligent Information Technologies for Industry" (IITI'17), Vol. 680, pp. 386-395, 2017.
[13] M. Yordanova, M. Vasileva, and R. Dimitrova, "Analysis of the mesh voltage calculation method in the presence of a two-layer soil," Proceedings of XLVIII-th International Scientific Conference on Information, Communication and Energy Systems and Technologies, ICEST'2013, vol. 2, pp. 723-726, 2013.
[14] E. Barudov, E. Panov, and S. Barudov, "Analysis of the electrical processes in a discrete alternating voltage regulator with active-inductive load," Annual Proceedings of the Technical University of Varna, vol. II, pp. 30-35, 2010. ISSN: 1311-896X. (in Bulgarian)
[15] E. Barudov, E. Panov, and S. Barudov, "Commutation processes at reactive loads in discrete alternating voltage regulator," Annual of the Technical University of Varna, vol. I, pp. 3-8, 2011. ISSN: 1311-896X. (in Bulgarian)

Exploration of the Efficiency of Autotransformer Discrete Alternating Voltage Regulators

Milena Ivanova
Department of Electric Power Engineering
Technical University of Varna
Varna, Bulgaria
m.dicheva@tu-varna.bg

Emil Panov
Department of Theoretical Electrical Engineering and Instrumentation
Technical University of Varna
Varna, Bulgaria
eipanov@yahoo.com

Emil Barudov
Department of Electrical Engineering
Nikola Vaptsarov Naval Academy
Varna, Bulgaria
ugl@abv.bg

(Published on the IEEE International Conference Automatics and Informatics ICAI'2020, 1-3 October 2020, Varna, Bulgaria, 5 pages. SBN 978-1-7281-9308-3, DOI: 10.1109/ICAI50593.2020.9311348.)

***Abstract*—In recent years, energy efficiency issues have become increasingly popular. This is due to the increase of both the electricity consumption and the price of the basic energy sources. One of the approaches for satisfaction the increased requirements for the quality of electrical power is related to the regulation of the amplitude of the input voltage, which changes over time. An effective method for limiting the supply voltage range can be realized by using autotransformer alternating voltage regulators. The paper is dedicated to the research on the loading of such regulators at different ranges of commutation (different input voltages) and the change of the efficiency coefficient. An automated computer program has been created, which allows the simulation of the processes in the discrete regulators in the MATLAB environment. An experimental check of a prototype was performed showing a good compliance with the simulation results.**

Keywords—autotransformer discrete alternating voltage regulator, efficiency coefficient, thyristor, commutation process

I. Introduction

In the recent years energy efficiency issues have become increasingly popular. This is due to the increase of the electricity consumption and the price of the basic energy sources. There is a great interest in more precise control of the parameters of the supplied electricity for industrial and domestic consumption [1, 2, 3].

One of the modern approaches for improving the quality of electricity is by adjusting the amplitude of the supply voltage. This type of the supply voltage control can be realized by the use of autotransformer discrete alternating voltage regulators (ADAVR) with semiconductor commutation elements (SCE) [4, 5, 6, 7, 8, 9, 10, 11].

Recently, a non-linear model of these regulators has been developed to help the study of the complex processes taking place in these devices [12, 13, 14, 15, 16, 17].

A corresponding computer program AVTO in the MATLAB environment was created on the base of that model in order to automate the researches on ADAVR [14, 15, 16]. The developed computer program AVTO allows simulations of the ADAVR loading at switching of different voltage-adding elements and a visualization of the change of the efficiency coefficient. This allows correct dimensioning of the size and number of the regulating ranges of ADAVR [13, 15].

The paper is dedicated to the exploration of the loading of the ADAVR at different ranges of commutation and the change of the efficiency coefficient. An experimental check of a prototype was performed showing a good compliance with the data from the simulations.

The main contribution is that an analysis of an ADAVR with a similar structure is made in terms of efficiency when working with active load.

In Section II an analysis of the processes in the studied ADAVR with four levels of commutation and active load is presented. The specifics of the commutation process due to the usage of thyristor switches have been considered. The simulation results for the explored four stationary regimes, which are connected with the commutation of only one of the four switches of ADAVR, are given in Section III. Conclusions regarding the alteration of the output parameters and the efficiency of the regulator are shown in Section IV.

II. Analysis

The parameters of the electricity quality, which are valid in the Republic of Bulgaria, are regulated in BDS EN 50160:2010 "Voltage characteristics of electricity supplied by public electricity networks" [18].

In Table 1 the parameters of the electricity, which the electricity distribution companies are required to deliver to the end-users, are shown.

TABLE I. CHARACTERISTICS OF THE POWER SUPPLY OF PUBLIC ELECTRICITY NETWORKS IN BULGARIA ACCORDING [18]

№	Characteristics	Low voltage grids	Medium voltage grids
1.	Frequency	49,5 ÷ 50,5 Hz (for 99,5 % from an annual period) or 47 ÷ 52 Hz	
2.	Voltage deviation	U_n ± 10 % (for each period of 1 week, 95 % of the average RMS voltage value for 10 min.) U_n +10/-15 % (for each period of 1 week, all average RMS voltage values for 10 min.)	
3.	Quick changes of voltage	Less than 5 % U_n; changes up to 10 % U_n with short duration can occur several times a day under certain conditions. Flicker: P_{lt} ± 1 (for every period of one week)	Less than 4 % U_n; changes up to 6 % U_n with short duration can occur several times a day under certain conditions. Flicker: P_{lt} ± 1 (for 95 % of period of one week)
4.	Short-term outages	Indications: from several tens to several hundreds	

The limitation of the range of the supply voltage amplitude change is realized by an automatic switching of the autotransformer (ATr) terminals. The terminals of ATr are positioned at the side of the load. In Fig. 1 the equivalent circuit of ADAVR with four SCE is presented. The scheme includes the parameters of the magnetic circuit, the commutating groups, the switch-off assemblies, the separate sections of the winding and the existing non-linear effects.

The model allows the exploration of the random commutations over time of consecutively switched neighboring SCE.

The use of thyristor switches leads to a specific type of a commutation process, which includes three separate intervals [14, 15, 16].

In the first interval ADAVR operates in stationary AC regime and it is described by loop equations in phasor form. Here, only one switch in ADAVR is closed.

The main form of the set of equations assumes the following phasor matrix form:

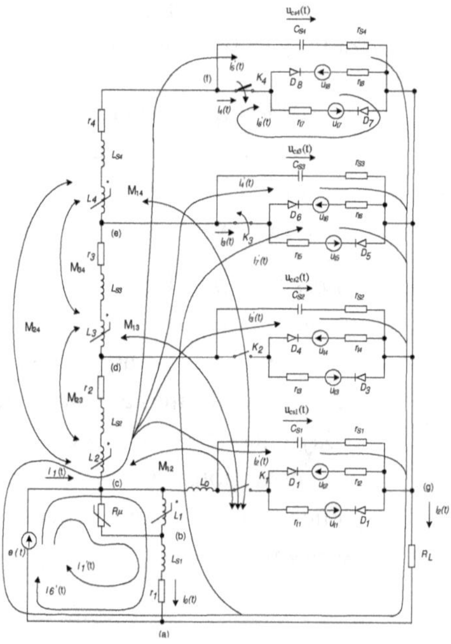

Fig. 1. Equivalent circuit of ADAVR.

$$[Y_I]\cdot[\dot{I}']=[\dot{E}] \tag{1}$$

where $[Y_I]$ is a contour impedance matrix of ADAVR; $[\dot{I}']$ is a matrix-column of the loop currents and $[\dot{E}]$ is a matrix-column of the contour electromotive forces (emf).

During the second interval, the first transient process in ADAVR is observed while two of the switches are closed. The analysis is made by the state variables method in time domain, where some of the state variables are loop currents, and the rest of them are voltages across the switch-off capacitors:

$$[A_2]\frac{d}{dt}[x(t)] = [B_2]\cdot[x(t)] + [e_2(t)] \tag{2}$$

During the third interval, the second transient process is observed and there only a single switch is closed. The analysis is done by the state variables method in the time domain. The number of state variables is already less than in the previous interval. The set of equations in the third interval is as follows:

$$[A_3]\frac{d}{dt}[x(t)] = [B_3]\cdot[x(t)] + [e_3(t)] \tag{3}$$

A computer program AVTO was created for simulating the processes in ADAVR with four SCE in the environment of MATLAB in view of its ability for mathematical calculations in matrix form and its capability to visualize the obtained numerical results in a graphical form. The computer program AVTO includes an input block for introducing the parameters of the computational process and the explored ADAVR. In this block the durations of the first and the third interval of the commutation process, the size of the calculation step, the effective value of the supply voltage and the parameters of the windings, the ferromagnetic core, the commutating elements, the switch-off assemblies and the load are set. The program calculates and visualizes the accompanying parameters of the commutation process.

The computer program AVTO is able to calculate the efficiency coefficient of the regulator η. It does not depend on the RMS value of the supply voltage E of ADAVR, but depends on the parameters of the regulator and the load resistance R_L. Really it depends on the input real power and the output real power of the regulator:

$$\eta = \frac{P_2}{P_{gen}}\cdot 100\% \tag{4}$$

III. Results

From the graphics (Fig. 2 ÷ Fig. 18), presented for the four operating cases of the regulator associated with the situation, when one of the four switches K_1, K_2, K_3 or K_4 is closed, it can be easily seen the dependencies of the values of the output voltage U_2, the real power at the input of ADAVR P_{gen}, the reactive power at the input of ADAVR Q_{gen}, the real power at the output of ADAVR P_2 and the efficiency coefficient η as functions of the output current I_2.

The four stationary regimes of the regulator, which were explored, are connected with the commutation of only one of the four switches of ADAVR:

A. First case: switch K_1 *is closed, and all other switches are opened.*

For that case the following graphical inscriptions are used (Fig. 2, Fig. 3 and Fig. 4):

when $E = 220V$ – symbol "star";

when $E = 200.1V$ – symbol "circle".

When the supply voltage E varies from 220V to 200,1V, the output voltage U_2 equals the input voltage of ADAVR E (i.e. it repeats E by value for each value of the load current I_2). The input reactive power of the regulator Q_{gen} is a constant at a given value of E in the range from 220V to 200,1V, i.e. it does not depend on load current I_2. Q_{gen} depends only on the value of the input voltage E.

When $E = 220V$, $Q_{gen} = 210.05VAr$ for the whole range of I_2, but when $E = 200,1V$, Q_{gen} decreases to 173,6VAr for all values of I_2, i.e. it drops down together with the value of E.

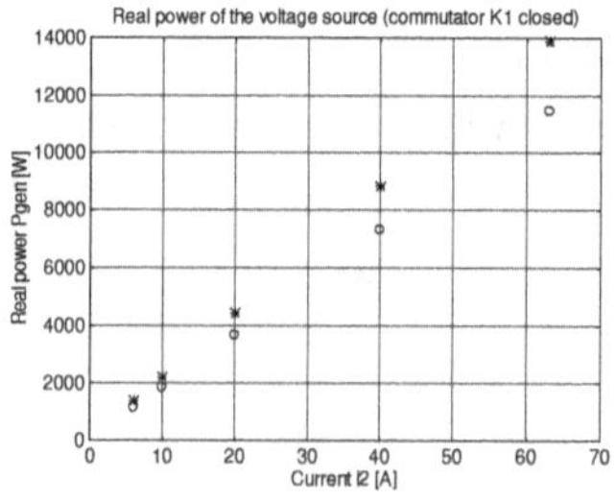

Fig. 2. Input real power P_{gen} versus the output current I_2 (K_1 – closed).

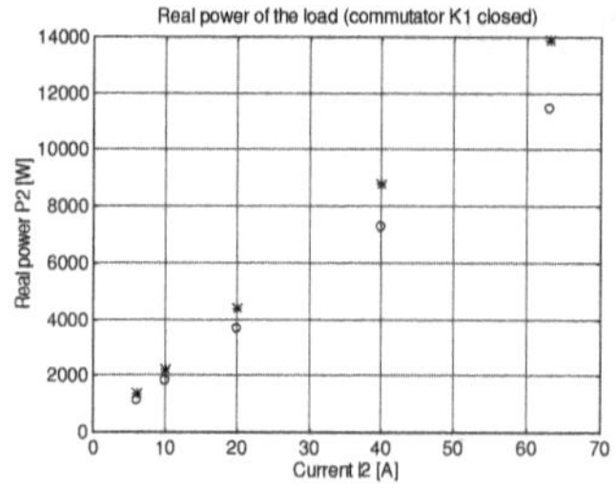

Fig. 3. Output real power P_2 versus the output current I_2 (K_1 – closed).

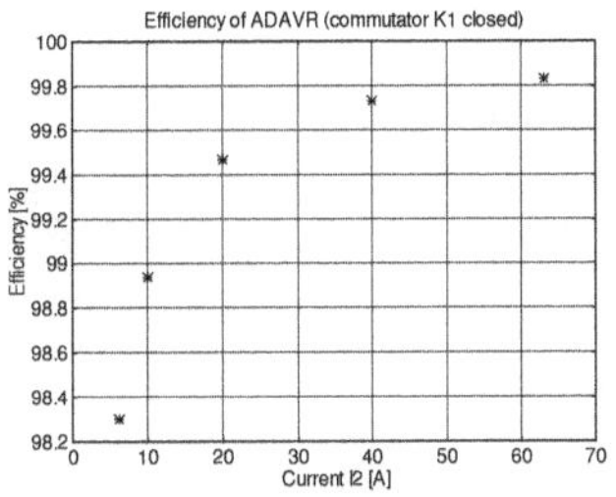

Fig. 4. Efficiency coefficient η versus the output current I_2 (K_1 – closed).

It is seen in Fig. 2 ÷ Fig. 4, that the input real power P_{gen}, the output real power P_2 and the efficiency coefficient η increase in the range of the current I_2. The dependency of the efficiency coefficient η on the values of I_2 does not change at different values of the supply voltage E within the range between 220V and 200,1V.

B. Second case: switch K_2 *is closed, and all other switches are opened.*

For that case the following graphical inscriptions are used (Fig. 5 ÷ Fig. 8):

when E = 200V – symbol "star";

when E = 180,1V – symbol "circle".

For each value of the supply voltage E in the range from 200V to 180,1V the output voltage U_2 remains almost constant and it does not depend on the value of the load current I_2.

When E = 200V and the load current has a value I_L = 6,19A, the output voltage is U_2 = 219,62V; but when E = 200V and I_2 = 63A, then U_2 = 219,1V, i.e. U_2 is nearly the same in the range of the output current I_2.

When E = 180,1V and the load current has a value I_L = 6,19A, the output voltage is U_2 = 197,66V; but when E = 180,1V and I_2 = 63A, then U_2 = 197,1V, i.e. U_2 is nearly the same in the range of the output current I_2 as in the previous case. It is seen in Fig. 5, Fig. 6 and Fig. 7, that the input real power P_{gen}, the input reactive power Q_{gen} and the output real power P_2 increase with the increase of the value of the output current I_2. It is also seen in Fig. 8, that the efficiency coefficient η has a maximum in the range of the current I_2. The dependency of the efficiency coefficient η on the values of the load current I_2 does not change at different values of the input voltage E within the range between 200V and 180,1V.

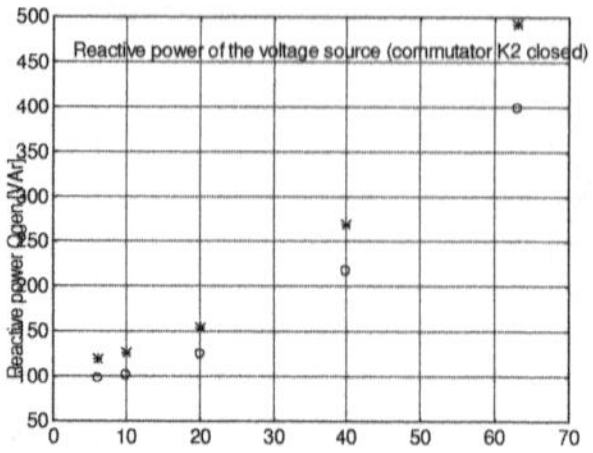

Fig. 5. Input real power P_{gen} versus the output current I_2 (K_2 – closed).

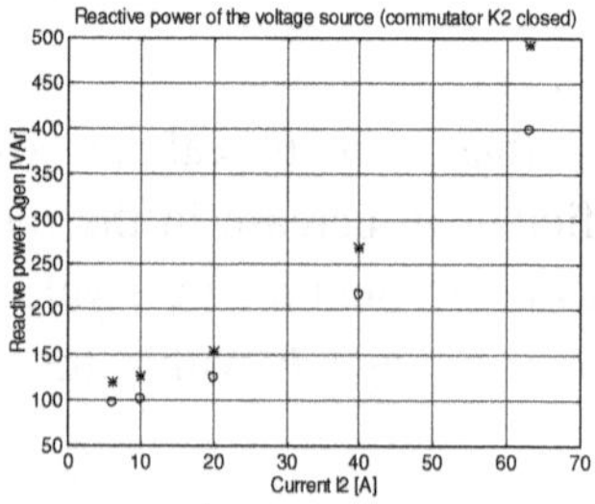

Fig. 6. Input reactive power Q_{gen} versus the output current I_2 (K_2 – closed).

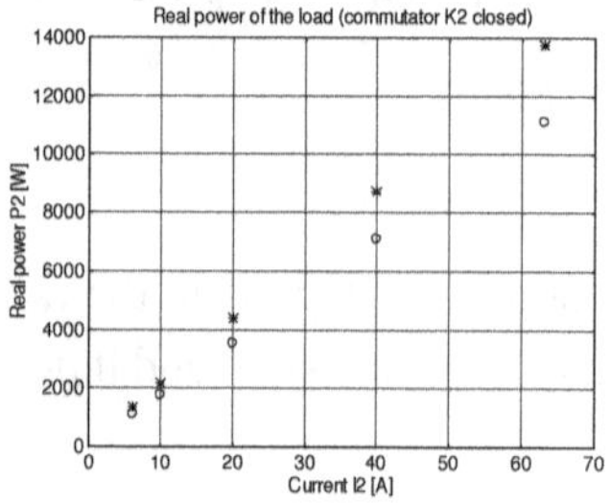

Fig. 7. Output real power P_2 versus the output current I_2 (K_2 – closed).

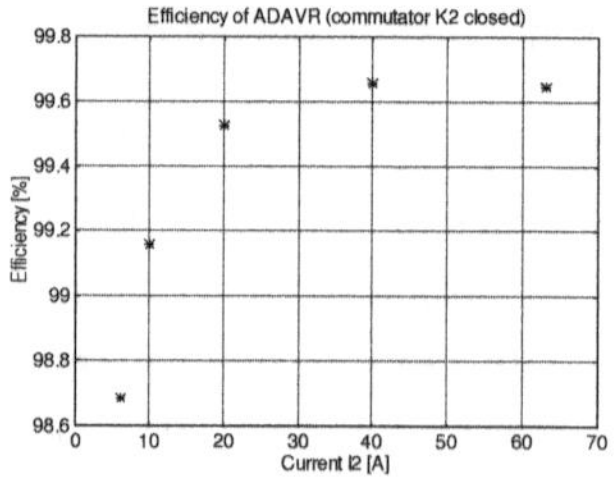

Fig. 8. Efficiency coefficient η versus the output current I_2 (K_2 – closed).

C. Third case: switch K_3 *is closed, and all other switches are opened.*

For that case the following graphical inscriptions are used (Fig. 9 ÷ Fig. 13):

when E = 180V – symbol “star”;

when E = 160,1V – symbol “circle”.

It is seen in Fig. 9, that the output voltage U_2 slightly decreases with the increase of the load current I_2. In Fig. 10 ÷ Fig. 12 the input real power P_{gen}, the input reactive power Q_{gen} and the output real power P_2 increase again with the increase of the value of I_2. The efficiency coefficient η has a maximum in the range of the output current I_2 (Fig.13). The dependency of the efficiency coefficient η on the current I_2 does not change at different values of the supply voltage E in the range between 180V and 160,1V.

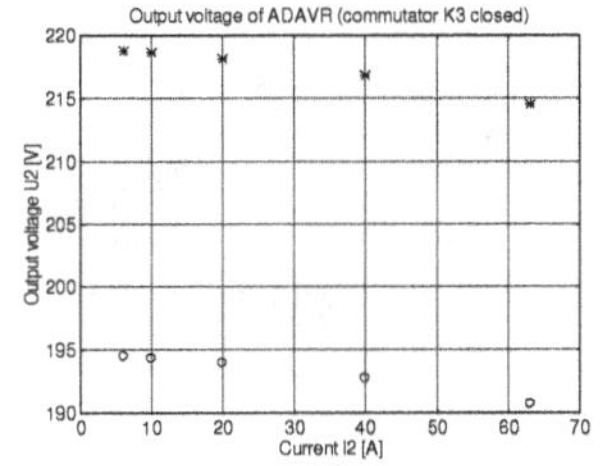

Fig. 9. Output voltage U_2 versus the output current I_2 (K_3 – closed).

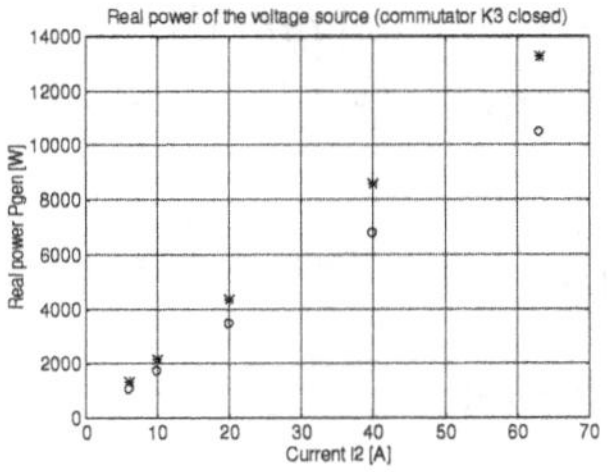

Fig. 10. Input real power P_{gen} versus the output current I_2 (K_3 – closed).

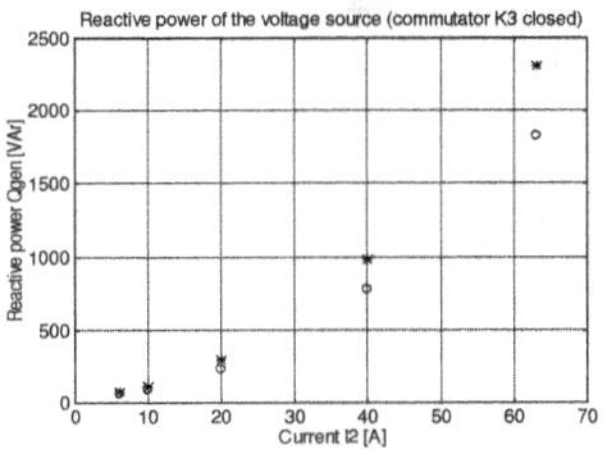

Fig. 11. Input reactive power Q_{gen} versus the output current I_2 (K_3 – closed).

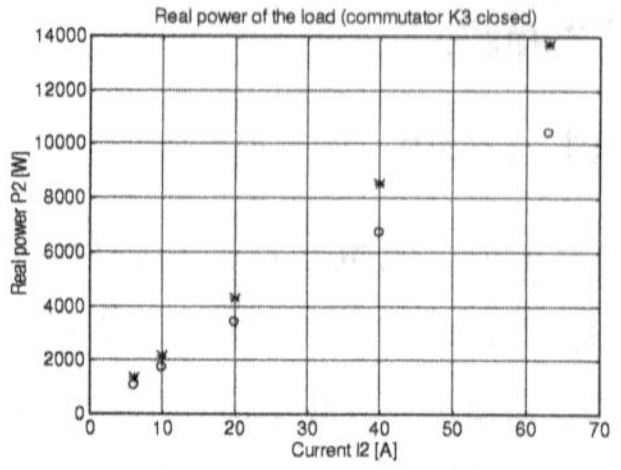

Fig. 12. Output real power P_2 versus the output current I_2 (K_3 – closed).

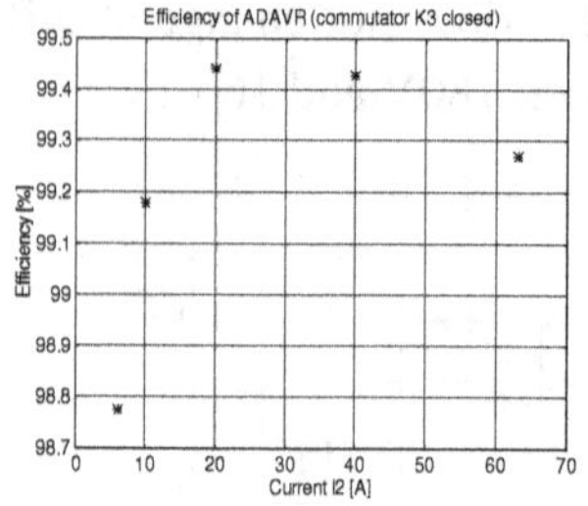

Fig. 13. Efficiency coefficient η versus the output current I_2 (K_3 – closed).

D. Fourth case: switch K_4 *is closed, and all other switches are opened.*

For that case the following graphical inscriptions are used (Fig. 14 ÷ Fig. 18):

when E = 160V – symbol "star";

when E = 150V – symbol "circle".

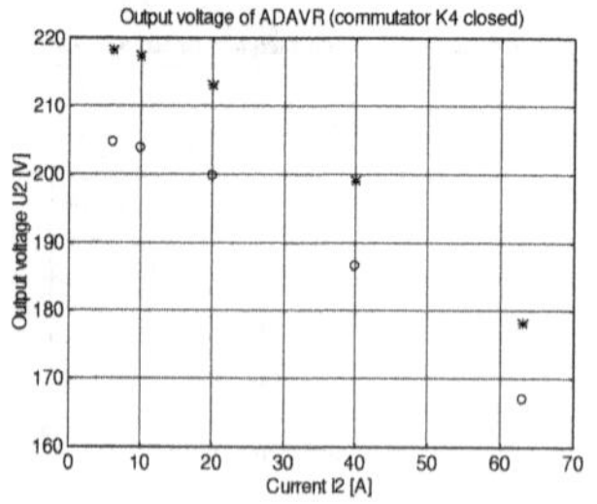

Fig. 14. Output voltage U_2 versus the output current I_2 (K_4 – closed).

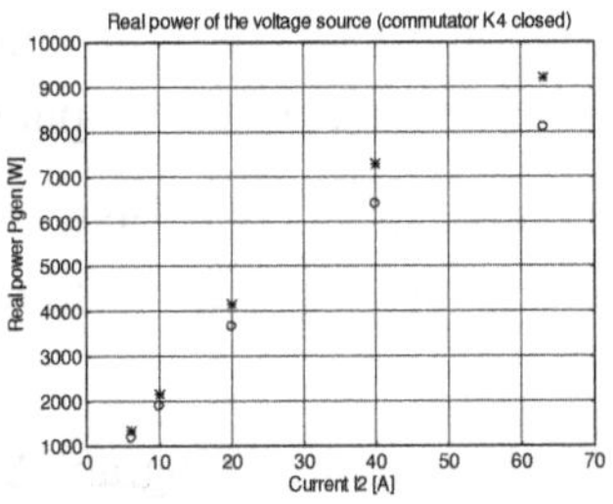

Fig. 15. Input real power P_{gen} versus the output current I_2 (K_4 – closed).

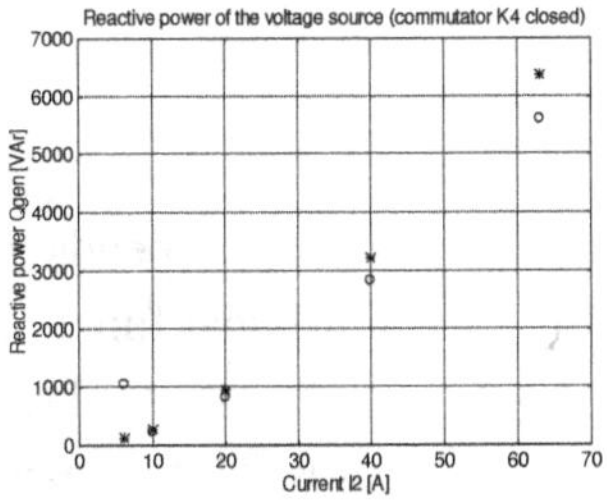

Fig. 16. Input reactive power Q_{gen} versus the output current I_2 (K_4 – closed).

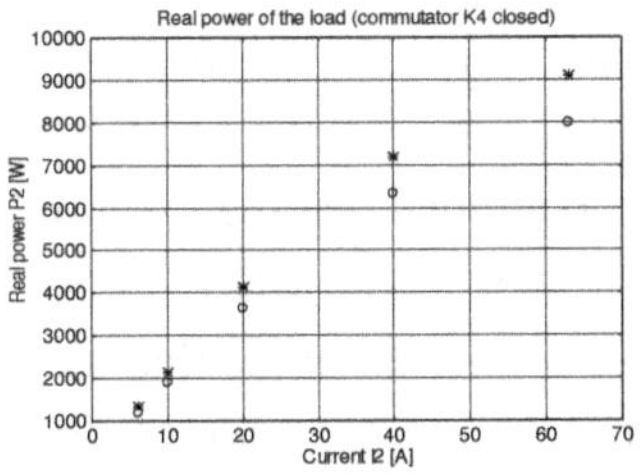

Fig. 17. Output real power P_2 versus the output current I_2 (K_4 – closed).

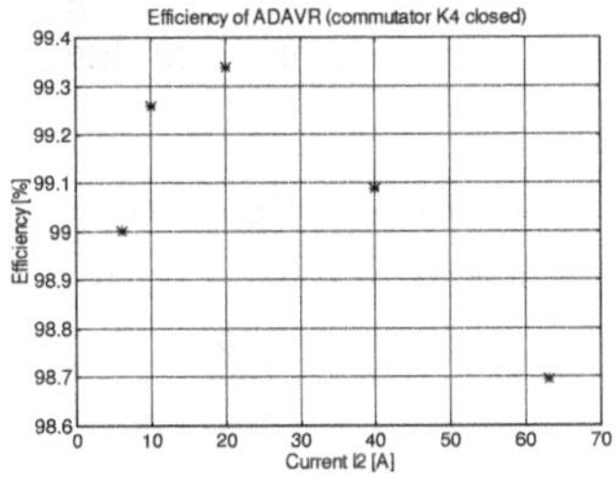

Fig. 18. Efficiency coefficient η versus the output current I_2 (K_4 – closed).

It is seen in Fig. 14, that the output voltage U_2 quickly decreases with the increase of the load current I_2. In Fig. 15, Fig. 16 and Fig. 17 the input real power P_{gen}, the input reactive power Q_{gen} and the output real power P_2 increase in the range of the output current I_2. It is also seen in Fig. 18, that the efficiency coefficient η has a maximum in the range of the output current I_2.

The dependency of the efficiency coefficient η on the values of I_2 does change at different values of the supply voltage E within the range between 160V and 150V.

The results, received by the computer program AVTO after exploration on the efficiency of ADAVR operating at different regimes, were experimentally checked and they showed a very good compliance with the data from the simulated processes.

IV. Conclusions

1. The efficiency coefficient η by theory depends on the parameters of the regulator's elements, and they are non-linear functions of the supply voltage E, the input current I_1, the current in the transverse branch I_0 of the regulator and the load current I_2. For the ADAVR of the explored type the efficiency coefficient η varies from 98,3% to 99,83%, which is an indicator for high quality of the studied ADAVR.
2. Having in mind the characteristics of the output electrical power of the regulator and the behavior of the output voltage U_2 with the change of the output current I_2 it can be concluded that the explored ADAVR fully covers the requirements of the Bulgarian standard BDS EN 50160:2010.
3. It is very well seen from the simulation results, that if the supply voltage E is smaller than 180,1V, and the output current I_2 is higher than 40 A, the consumed reactive power Q_{gen} becomes a factor, which cannot be neglected.
4. The extremely high energy efficiency (up to 99,83%) of the studied ADAVR makes it very suitable for mass use in practice due to its good performance.

Acknowledgment

The current scientific work is financed by the National Program "Young Scientists and Postdoctoral Researchers", 2020.

References

[1] W. Kazibwe, and M. Senduala, Electric Power Quality Control Techniques. N. Y., Van Nostrand Reinhold, 1993.

[2] M. Bollen, Understanding Power Quality Problems: Voltage Sags and Interaptions. N. Y., IEEE Press, 2000.

[3] P. Prodanov, and D. Dankov, "Reliability of power supplies for induction heating through an analysis of the states in operating modes," XIX^{-th} International

Symposium on Power Electronics, Ee, pp. 1-5, 2017, DOI:10.1109/PEE.2017.8171671.
[4] G. Goedde, L. Kojovic, and E. Knabe, “Overvoltage protection for distribution and low-voltage equipment experiencing sustained overvoltages,” IEEE Power Engineering Society. 1999 Winter Meeting, New York, USA, 1999, Print ISBN: 0-7803-4893-1, DOI: 10.1109/PESW.1999.747379.
[5] J. Harlow, “Transformers” in The Electric Power Engineering Handbook. Ed. L.L Grigsby Boca Raton: CRC Press LLC, 2001.
[6] R. Tillman, “Loading Power Transformes” in The Electric Power Engineering Handbook. Ed. L. L. Grigsby, CRC Press, Boca Raton, FL, 2001.
[7] M. Bishop, J. Foster, and D. Down, “The application of single-phase voltage regulators on three-phase distribution systems,” IEEE Industry Application Magazine, pp. 38-44, July/August 1996.
[8] H. Hart, and R. Kakalec, “The derivation and application of design equations for ferroresonant voltage regulators and regulated rectifiers,” IEEE Trans. on Magnetics, vol. 7, pp. 205-211, 1971.
[9] M. Huang, J. Lu, and Y. Peng, “Research on trigger methods for thyristor ac-voltage regulator,” Journal on Power Electronics, vol. 2, pp. 54-55, 2004.
[10] M. Hoque, and A. Mahmod, “An improved automatic voltage regulation system with apposite hysteresis and immense precision,” Chittagong University Journal of Science, vol. 33, pp. 21-33, 2010.
[11] N. Kutkut, R. Schneider, T. Grant, and D. Divan, “AC voltage regulation technologies,” Power Quality Assurance, pp. 92 -97, 1997.
[12] E. Barudov, S. Barudov, and E. Panov, “Study of the Transient Process Length in Step Voltage Regulator”, Journal “Acta Universitatis Pontica Euxinus,” vol. 3, n. 1, pp. 91÷96, 2004, ISSN 1312-1669.
[13] S. Barudov, and E. Barudov, Discrete Alternating Current Regulators and Stabilizers. Sofia-Moscow, Pensoft, 2006, ISBN 978-954-642-276-7.
[14] E. Barudov, E. Panov, and S. Barudov, “Analysis of electrical processes in alternating voltage control systems,” XII-th International Symposium “Materials, Methods&Technologies (MMT), 11-15 June 2010, Sunny Beach, Bulgaria, published in Journal of International Scientific Publications: Materials, Methods & Technologies, vol. 4, part 1, pp. 154 – 182, 2010, ISSN: 1313 2539, http://www.science-journals.eu.
[15] E. Panov, E. Barudov, and S. Barudov, “Vector analysis and comparative valuation of precise and approximate non-linear models of discrete regulator with reducing input AC voltage,” Proceedings of XLVIII-th International Scientific Conference on Information, Communication and Energy Systems and

Technologies ICEST'2013, Ohrid, Macedonia, vol. 2, pp. 771-774, 26-29 June 2013, ISBN 978-9989-786-89-1.
[16] E. Panov, E. Barudov, and M. Ivanova, "Exploration of the electric processes in discrete alternating step-voltage regulators," Proceedings of the XX-th International Symposium on Electrical Apparatus and Technologies SIELA'2018, Bourgas, Bulgaria, pp. 325 – 328, 2018, ISBN 978-1-5386-3418-9.
[17] S. Barudov, E. Panov, E. Barudov, and M. Ivanova, Discrete Stabilizer of AC Voltage, Patent BG1727 U1/07.08.2013. (in Bulgarian)
[18] BDS EN 50160:2010/A1:2015 "Voltage characteristics of electricity supplied by public electricity networks".

Chapter 9. Protection Regimes in the ADAVR

Protection Regimes and Electrical Quantities in Autotransformer Alternating Voltage Regulators

Emil Panov
Department of Theoretical Electrical Engineering and Instrumentation
Technical University - Varna
Varna, Bulgaria
e-mail: eipanov@yahoo.com

Dimitar Dimitrov
Department of Electrical Engineering and Electrotechnology
Technical University - Varna
Varna, Bulgaria
e-mail: prof_dimitrov@abv.bg

Emil Barudov
Department of Electrical Engineering,
Nikola Vaptsarov Naval Academy,
Varna, Bulgaria,
e-mail: ugl@abv.bg

Milena Ivanova
Department of Electric Power Engineering
Technical University - Varna
Varna, Bulgaria
e-mail: m.dicheva@tu-varna.bg

(Published in the Proceedings of the XXI-th International Symposium on Electrical Aparatus and Technologies SIELA'2020, 3-6 June 2020, Bourgas, Bulgaria, pp. 338-341, Electronic ISBN: 978-1-7281-4346-0, USB ISBN: 978-1-7281-4345-3, Print on Demand (PoD) ISBN: 978-1-7281-4347-7, DOI: 10.1109/SIELA49118.2020.9167144.)

***Abstract*—The paper presents an analysis of the protection regimes of an autotransformer alternating voltage regulator with four thyristor switches**

by using overcurrent protections at the input, at the output of the device and before each of the commutating elements of the regulator. Each commutation element (thyristor switch) of the regulator is also protected by an overvoltage protection, which can be an electronic device or a *RC* group. The electrical quantities (the input and the output currents and voltages) of the regulator with the considered protections have been determined by using a previously developed MATLAB simulation model of the device. The simulation results were experimentally verified and good coincidence was achieved.

Keywords—autotransformer discrete alternating voltage regulator, overcurrent protection, overvoltage protection, thyristor switches.

I. Introduction

Nowadays, voltage regulation is necessary for protecting industrial and household consumers against voltage fluctuations in the power supply, because they can cause severe damages to the equipment. Most electrical devices can stand supply voltage up to a certain limit. The maintaining of a constant voltage level at the input of the consumer's device can be achieved by using ferroresonant regulators [1], thyristor alternating voltage regulators [2], electronic regulators [3, 4] or autotransformer discrete alternating voltage regulators [5].

The previous work of the authors is related to the investigation of autotransformer discrete alternating voltage regulators (ADAVR) [5, 6, 7]. The design of such regulators requires the usage of modern approaches for computer analysis and simulation of the complex processes occurring in them [8, 9]. Another important issue is that the device itself must be protected against overcurrents and overvoltages.

The main objective of the present research is to study the processes in ADAVR at emergency regimes, which are protected against overcurrents and overvoltages. The specific processes have been explored by using a MATLAB simulation model, previously developed by the authors [6, 10]. The novelty of the present work is the combination of the model of ADAVR with different types of protections and the analysis of the operation of the protected device. Finally, an experimental verification of the simulation results has been done.

II. Analysis

The equivalent circuit of the studied ADAVR is shown in Fig. 1. It includes four semiconductor commutating elements (thyristor switches). The operating voltage range of the regulator is from *160V* to *220V*. The parameters of the magnetic circuit, the commutation groups (Fig. 2), the switch-off assemblies and the parameters of the separate sections of the winding have been considered in the model [7]. The non-linearities of the elements have been also taken into account. The created model enables detection of the random commutations of adjacent tryristor switches in the time domain.

The overcurrent protections (OCP) are implemented at the input, at the output and before each of the commutating elements of the regulator. The scheme of a fast-acting electronic OCP is shown in Fig. 3. The simplest option is to use a fast-acting fuse.

The semiconductor thyristor switches ($S_1 \div S_4$) (Fig. 2) are protected by overvoltage protections (OVP). Different variants of an OVP are presented in Fig. 4 and Fig. 5, as for the current case the protection is realized by RC groups (snubber circuit thyristor overvoltage protection - Fig. 5). The two types of protections that are used in the model react to the rms values of the electrical quantities (currents or voltages).

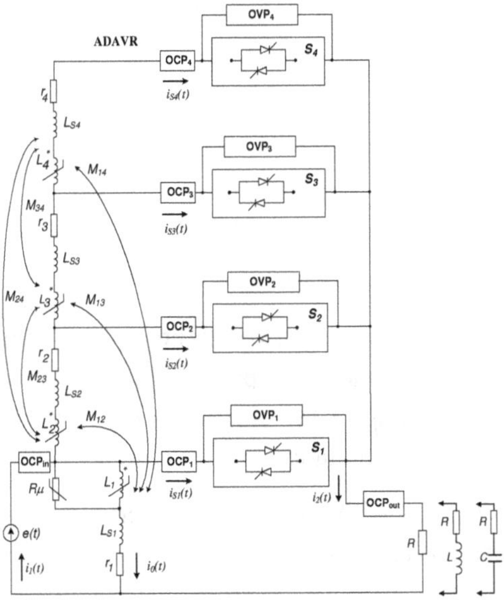

Fig. 1. Equivalent circuit of ADAVR with overcurrent and overvoltage protections.

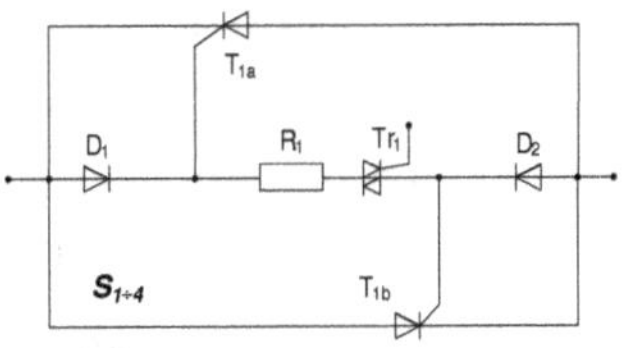

Fig. 2. Schematic diagram of a thyristor switch ($S_1 \div S_4$).

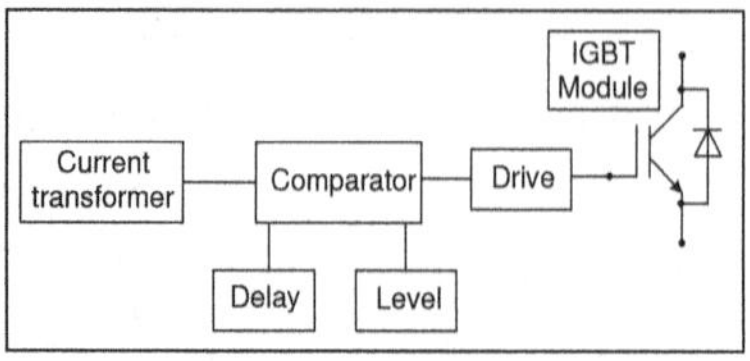

Fig. 3. Fast-acting electronic overcurrent protection module.

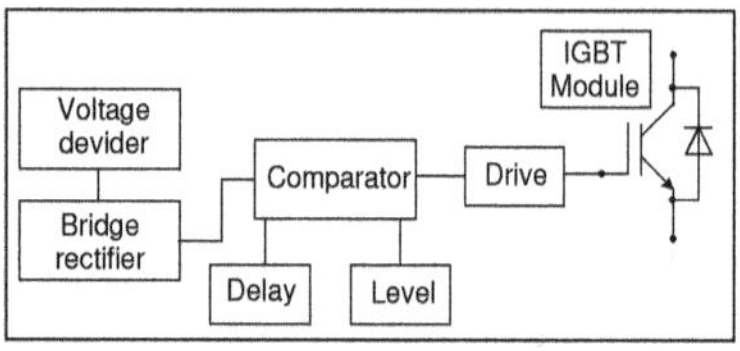

Fig.4. Fast-acting electronic overvoltage protection module.

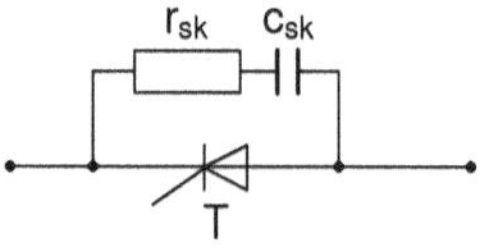

Fig. 5. Snubber circuit thyristor protection (OVP).

The characteristic states of the model are:

1) When U_I=E = *const.* – only one switch is closed;

2) When $U_{I\,(t+0)}>U_{I\,(t-0)}$ – there are two subintervals - a) when two switches are simultaneously closed and connected in parallel. In that case there is a short circuit of one section of the winding of the autotransformer; b) when only the switch from the newly switched commutation level is closed;

3) When $U_{I(t+0)}<U_{I(t-0)}$ - with the same subintervals as in the previous case. During the first subinterval the commutation transient processes are very fast. Therefore, the previously working thyristor commutator is switched off very quickly and it can't cause flowing of commutation current, which can damage the switching elements.

The analysis of the electrical quantities (the input and the output currents and voltages) in different protection regimes (at the input, at the output and in the circuits around the thyristor commutation switches) is done by using the computer program AVTO, developed in MATLAB [6, 7, 9, 10].

Three intervals of the most dangerous commutation processes in ADAVR were analyzed:

1) When the ADAVR operates in stationary ac regime only one switch (S_3 or S_4) in it is closed. The electric equilibrium is described by the loop currents method in phasor form. The following phasor matrix form is used:

$$[A_1].\left[\dot{X}_1\right]=\left[\dot{F}_1\right] \tag{1}$$

where $[A_1]$ is a square matrix of the system including r, L, M and C elements, $\left[\dot{X}_1\right]$ is a matrix-column including the phasors of the loop currents $\left[\dot{I}_1\right]$ and $\left[\dot{F}_1\right]$ is a matrix-column of the input excitation.

The number of the loop currents depends on the type of the load. For R and RC load there are seven loop currents and for RL load the loop currents are eight. The loops for the analysis are unique for each time interval. The matrix $[A_1]$ is non-singular (its determinant is nonzero) even at short circuit at the output – $Z_L=0\Omega$.

2) The first transient process in the ADAVR occurs provided the switches S_3 and S_4 are closed.

The general matrix form of the system of equations is determined by applying the state variable approach (2):

$$[A_2].\frac{d}{dt}\left[X_2(t)\right]=[B_2].\left[X_2(t)\right]+[C_2].\left[f_2(t)\right] \tag{2}$$

where $[A_2]$ is a matrix of the system consisting of L, M and C elements. It is a highly sparse matrix which is singular, i.e. its determinant is very close to zero; $[B_2]$ is a matrix consisting of the active resistances in the system – the resistances of the autotransformer winding and load resistance; $[X_2(t)]$ is a matrix-column including eight loop currents for R and RC load or nine loop currents for RL load and four capacitor voltages; $[C_2]$ is a matrix containing zero and unit elements. It is not influenced by the load; $[f_2(t)]$ is a matrix-column of the input excitation and the residual voltage drops over the triggered thyristors ($0{,}7V$).

3) The second transient process occurs, but here only the switch S_3 or S_4 is closed (3).

The general matrix form of the system of equations is determined by applying the state variable approach (3):

$$[A_3].\frac{d}{dt}\left[X_3(t)\right]=[B_3].\left[X_3(t)\right]+[C_3].\left[f_3(t)\right] \tag{3}$$

The description of the matrices is similar to that of the second interval. However, the loops for determining the loop currents are different. There are seven loop currents for R and RC load. And for RL load there are eight loop currents. The matrix $[A_3]$ does not change at short circuit at the output. It is also a singular matrix. The matrix $[B_3]$ changes at $Z_L=0\Omega$ as it still contains the resistances of the autotransformer winding. The load does not influence on the matrix $[C_3]$.

Summarizing the three intervals of the commutation process it can be noted that the matrices $[A_1]$, $[B_2]$ and $[B_3]$ change at different load impedances (short circuit) but this fact does not lead to failure of the computational process. Each of the matrices $[A_2]$, $[A_3]$, $[B_2]$ and $[B_3]$ consists of nine submatrices for the purpose of the calculation procedure. The values of the parameters and the configurations of the loops in the equivalent circuit from Fig. 1, set in the simulation model, are different for each of the three intervals of the commutation process. The values of the parameters are changing in the simulation program depending on the input voltage of ADAVR.

Iterative algorithm is used for solving the equations from the second and the third interval in the simulation model. Multi-step methods have been used such as the Runge-Kutta method, Adams-Bashforth method or Adams-Moulton method [10, 11, 12].

The calculation step of the iterative process is automatically adjusted until a certain accuracy of the final results is achieved.

III. Results

The electrical quantities of the studied ADAVR have been determined in the protection regimes corresponding to the second and the third characteristic state described in the previous section. The limit voltages at which a commutation of the different sections occurs are $160V$ (S_4), $180V$ (S_3), $200V$ (S_2) and $220V$ (S_1). Higher values of the current at the moment, when two neighboring thyristor switches are simultaneously closed, have been observed at lower values of the input voltage. Therefore, the results for the commutation at the lowest input voltages of the device's operating range are presented below.

In Fig. 6 and Fig. 7 the experimental oscillograms of the currents through S_3 and OCP_3 (respectively S_4 and OCP_4) are presented at *RL* load ($R=61{,}8\Omega$, $L=1{,}76$H, which are connected in parallel) and $U_1=159V$ (slowly increasing to $161V$) corresponding to the first subinterval of the second characteristic state. The commutation angle is $225°$. The upper curve on the oscillograms is the input voltage, while the lower curve presents the current $i_{s3}(t)$ or $i_{s4}(t)$. On the oscillograms in vertical direction $1mV/div$ corresponds to $1A/div$. The observed current peaks could reach $600A \div 700A$, which could damage the corresponding thyristor switches. At the same time there is a current peak of the input current $i_1(t)$ (Fig. 8), which is much smaller than the peaks of the previous two currents.

The oscillograms of the currents through S_3 and S_4, when the protection modules OCP_3 and OCP_4 are operating, are presented in Fig. 9a and Fig. 9b. The same curves are measured, when the input voltage decreases below *160V* and the model is in the third characteristics state. No current peaks can be observed in that case, which does not overload the switches.

The experimental results are compared with the simulation results and a very good coincidence is achieved as shown in Fig. 9c and Fig. 9d.

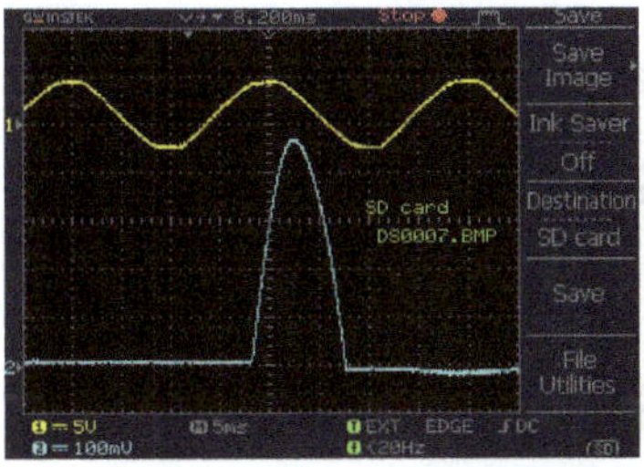

Fig. 6. Oscillogram of $i_{S3}(t)$.

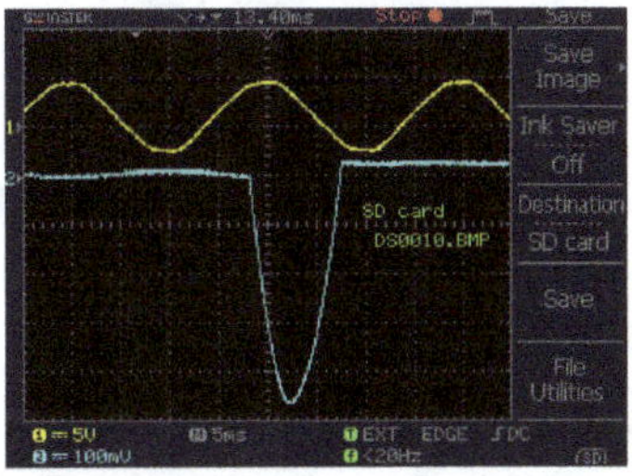

Fig. 7. Oscillogram of $i_{S4}(t)$.

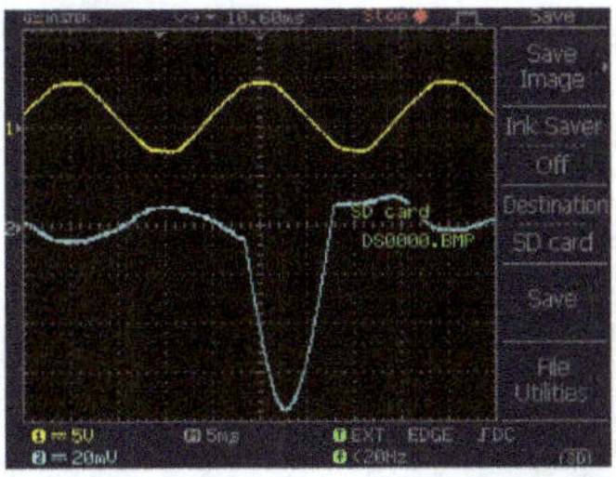

Fig.8. Oscillogram of $i_1(t)$.

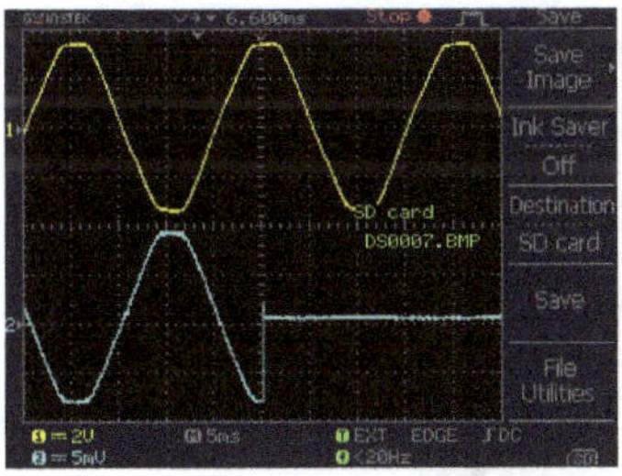

a) Oscillogram of $i_{S3}(t)$ (OCP_3).

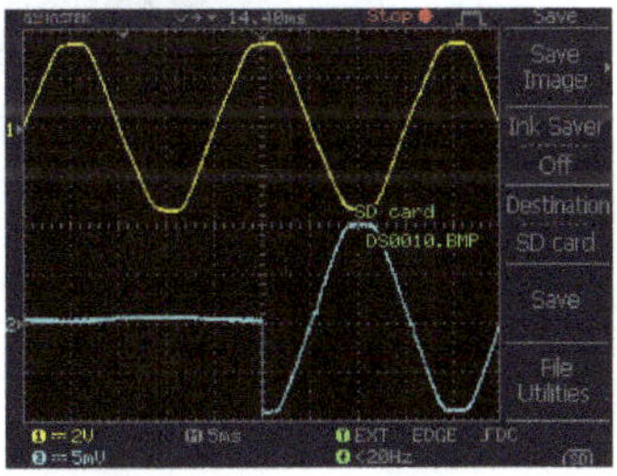

b) Oscillogram of $i_{S4}(t)$ (OCP_4).

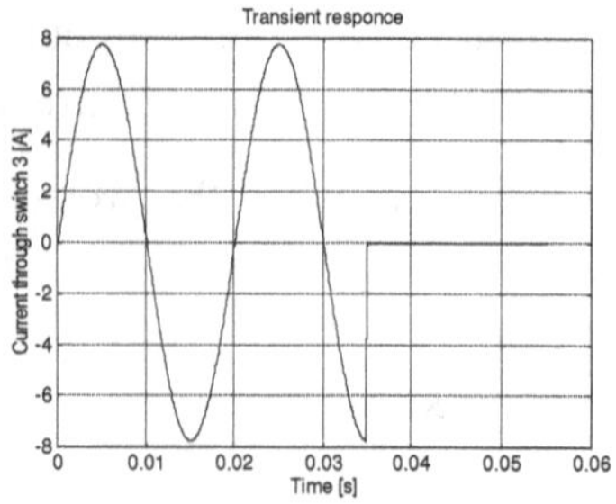

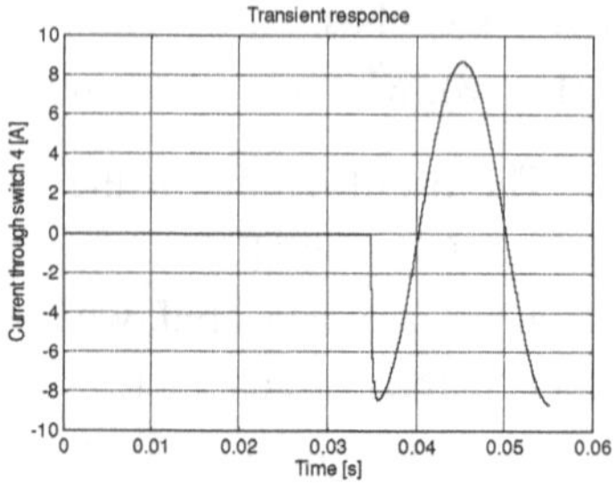

c) Computer simulation of $i_{S3}(t)$. d) Computer simulation of $i_{S4}(t)$.

Fig.9 Currents through the switches S_3 and S_4 at $U_1=E=160V$ (decreasing), $\varphi=270°$, $R=35,38\Omega$

The results for the output current at load parameters $R=35,7\Omega$, $L=1,76H$, commutation angle of $225°$ and input voltage $160V$, which is increasing, are shown in Fig. 10. The comparison of the results showed a good adequacy of the model (error less than 1%).

The results, derived by simulations, for the current and the voltage of the OVP_4 module at $R=30,23\Omega$, $L=2,095H$ and $U_1=160V$, when the model is analyzed in the third characteristics state (U_1 is decreasing), can be seen in Fig. 11. The observed periodically attenuating process is an evidence for the operation of the snubber circuit.

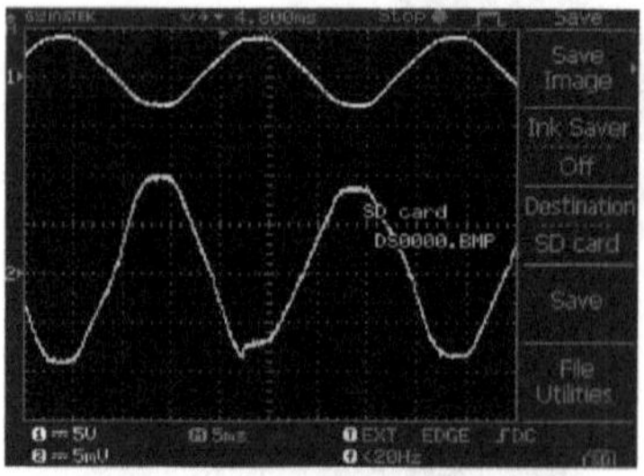

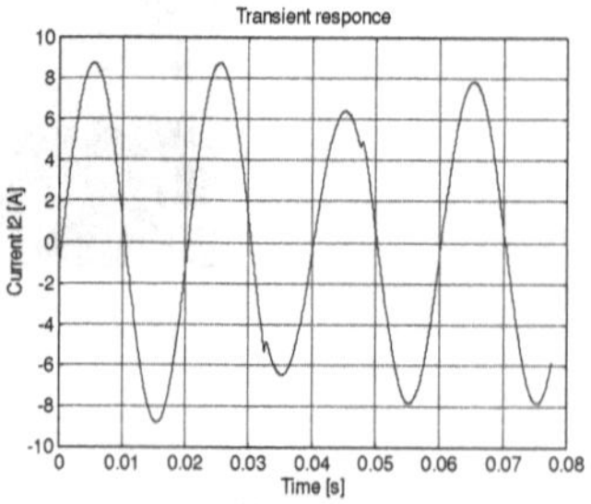

a) Experiment oscillogram. b) Computer simulation.

Fig.10 Output current $i_2(t)$ (OCP_{out}).

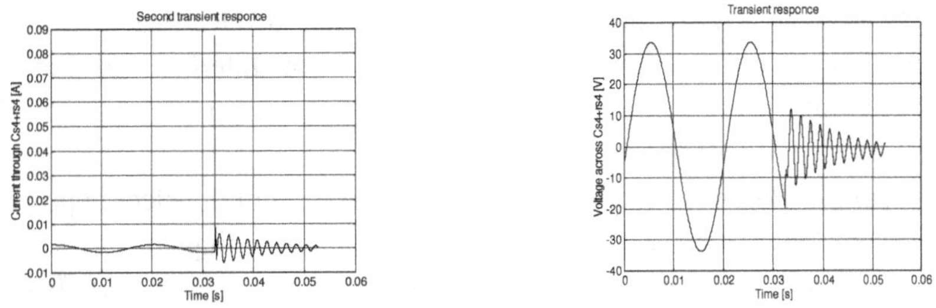

a) Current through OVP_4. *b) voltage over OVP_4.*

Fig. 11. Current and voltage in the snubber circuit.

IV. Conclusions

A simulation model of ADAVR with overcurrent and overvoltage protections including the non-linearities of all elements of the circuit was proposed and used for determining the electrical quantities of the protection devices.

In this way the currents flowing through the input, the output and through the semiconductor commutation elements of the explored ADAVR, as well as the transient processes in the different sections of the regulator (at the different commutation levels), were obtained.

The results from the simulations can be used for more precise setting of the different devices for protection of the explored ADAVR.

At the same time the experimental results showed a very good coincidence with these from the simulations (error of the model less than *1%*), which is a good verification of the validity of the obtained data.

References

[1] H. Hart and R. Kakalec, "The derivation and application of design equations for ferroresonant voltage regulators and regulated rectifiers," IEEE Trans. on Magnetics, vol. 7, pp. 205-211, 1971.

[2] M. Huang, J. Lu, and Y. Peng, "Research on trigger methods for thyristor ac-voltage regulator," Journal on Power Electronics, vol. 02, pp. 54-55, 2004.

[3] M. Hoque and A. Mahmod, "An improved automatic voltage regulation system with apposite hysteresis and immense precision," Chittagong University Journal of Science, vol. 33, pp. 21-33, 2010.

[4] N. Kutkut, R. Schneider, T. Grant, and D. Divan, "AC voltage regulation technologies," Power Quality Assurance, pp. 92 -97, 1997.

[5] S. Barudov and E. Barudov, Discrete Alternating Current Regulators and Stabilizers, Sofia-Moscow, Pensoft, 2006, 127 pages, ISBN 978-954-642-276-7.

[6] E. Barudov, E. Panov, and S. Barudov,"Analysis of electrical processes in alternating voltage control systems," Journal of International Scientific Publication: Materials, Methods & Technologies, vol. 4, part 1, pp. 154÷182, 2010, ISSN 1313-2539.

[7] E. Panov, E. Barudov, and M. Ivanova, "Exploration of the electric processes in discrete alternating step-voltage regulators," Proceedings of the XX-th Int. Symposium on Electrical Apparatus and Technologies (SIELA), Bourgas, Bulgaria, pp. 325-328, 2018, ISBN 978-1-5386-3418-9.

[8] V. Veselinov, Investigation of Emergency Processes in Power Electronic Circuits. Auto-essay on PhD dissertation, Technical University of Sofia, Sofia, 2011, 32 pages. (in Bulgarian)

[9] E. Panov, M. Ivanova, and E. Barudov, "Study of the electrical characteristics of autotransformer discrete alternating voltage regulators with R-L loads," Proceedings of the XVI-th International Conference on Electrical Machines, Drives and Power Systems (ELMA), Varna, Bulgaria, pp. 321-325, 2019, ISBN 978-1-7281-1412-5.

[10] E. Panov, E. Barudov, and S. Barudov, "Vector analysis and comparative valuation of precise and approximate non-linear models of discrete regulator with reducing input ac voltage," Proceedings of the XLVIII-th International Scientific Conference on Information, Communication and Energy Systems and Technologies (ICEST), Ohrid, Macedonia, vol. 2, pp. 771-774, 2013, ISBN 978-9989-786-89-1.

[11] V. Chauhana and P. Srivastavab, "Computational techniques based on Runge-Kutta method of various order and type for solving differential equations," International Journal of Mathematical, Engineering and Management Sciences vol. 4, n. 2, pp. 375–386, 2019, https://dx.doi.org/10.33889/IJMEMS.2019.4.2-030.

[12] J. Butcher, "Numerical methods for ordinary differential equations in the 20-th century," Journal of Computational and Applied Mathematics 125, pp. 1–29, 2000.

Chapter 10. Two-Port Parameters of the ADAVR

Exploration of the Two-Port Parameters of Autotransformer Discrete Alternating Voltage Regulators

Emil Panov
Department of Theoretical Electrical Engineering and Instrumentation
Technical University of Varna
Varna, Bulgaria
eipanov@yahoo.com

Emil Barudov
Department of Electrical Engineering
Nikola Vaptsarov Naval Academy
Varna, Bulgaria
ugl@abv.bg

Milena Ivanova
Department of Electric Power Engineering
Technical University of Varna
Varna, Bulgaria
m.dicheva@tu-varna.bg
(Published in South Florida Journal of Development, Miami, USA, v.3, n.6. p. 7049-7063, nov/dec., 2022. ISSN 2675-5459, DOI: 10.46932/sfjdv3n6-051.)

Abstract— **The present paper is dedicated to a new type of analysis of an autotransformer discrete alternating voltage regulator presented as a two-port with a load. The ABCD parameters of the transmission matrices were determined by using a simulation model of the regulator in the environment of MATLAB, which includes four semiconductor commutation elements (thyristor switches), corresponding to different supply voltages and considers the nonlinearities of the regulator's circuit. These ABCD parameters can be used for quick and accurate analysis of such a device in case of arbitrary type of the load – active, inductive or capacitive. The received results were experimentally verified in the case of an active load and a good coincidence less than 1% was achieved.**

Keywords—autotransformer discrete alternating voltage regulator, two-port, transmission matrix

I. Introduction

The autotransformer discrete alternating voltage regulators (ADAVR) are widely used in regulating the input supply voltage of household consumers in order to improve the quality of the electrical energy. The study of such regulators is a subject of number of scientific researches [1, 2, 3, 4, 5, 6, 7]. Their precise analysis and simulation are usually very difficult task. The purpose of the present study is to examine an ADAVR with four switching levels at different values of the input voltage as a two-port (TP) and to obtain its ABCD parameters and the specific systems of equations at arbitrary type of the load – active, inductive or capacitive. This new approach for research of such devices gives the opportunity for their quick and accurate analysis at any type of their load.

The scheme of the studied regulator, which includes four semiconductor commutating elements $S_1 \div S_4$ (thyristor switches), is shown in Fig. 1. The replacement circuit of the ADAVR as a TP is given in Fig. 2, where:

- The input terminals are (1) and (1');
- The output terminals are (2) and (2');
- $\dot{I}_1$ and $\dot{U}_1$ are the phasors of the input current and the input voltage;
- $\dot{I}_2$ and $\dot{U}_2$ are the phasors of the output current and the output voltage;
- $\dot{E}_1$ is the phasor of the supply voltage of the input source;
- Z_L is the impedance of the load.

The study is carried out in case of forward transmission of the TP.

For the investigated ADAVR the input and the output quantities (currents and voltages) had been determined at different loads and values of the supply voltage in previous researches by conducting simulations and experiments [8, 9, 10, 11, 12]. The simulations were performed in the environment of MATLAB by using a previously created nonlinear model by the authors [13, 14, 15, 16, 17, 18, 19]. The final results were compared to the experimental data and the correspondence is very good. The following peculiarities were taken into account in the study of the explored regulator:

- ferromagnetic core of the autotransformer;
- commutation groups;
- switch-off assemblies;
- separate sections of the winding.

The existing non-linearity caused by the ferromagnetic core of the autotransformer was also considered.

The commutation elements of ADAVR contain semiconductor elements (thyristors).

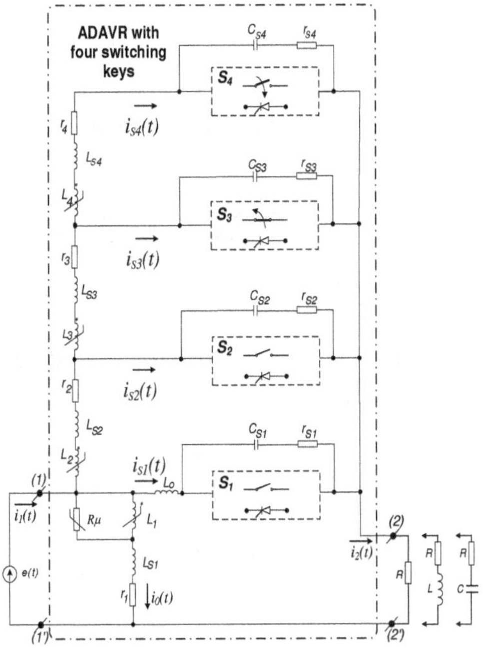

Fig. 1. Equivalent circuit of ADAVR with four levels of switching according to the values of the input supply voltage.

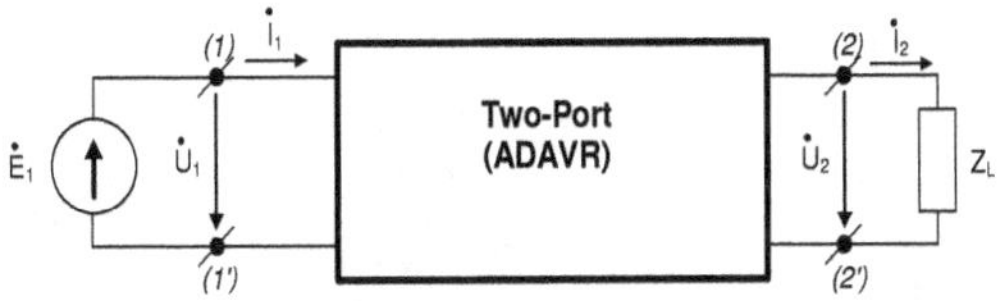

Fig. 2. Replacement circuit of an ADAVR explored as a TP at forward transmission.

II. Results and analysis

There is a great practical interest to determine the ABCD parameters of the transmission equations, which form the corresponding transmission system of equations (1) of the TP (the so-called A – system of equations) for the case of forward transmission:

$$\begin{cases} \dot{U}_1 = A\dot{U}_2 + B\dot{I}_2 \\ \dot{I}_1 = C\dot{U}_2 + D\dot{I}_2 \end{cases} \quad (1)$$

The system (1) allows at known values of the phasors of the output current and voltage to be determined the input current and voltage of the TP.

The ABCD parameters of the A-matrix can be calculated by the help of the input impedances at two different regimes: a) when there is an open circuit at the output terminals of the TP - $Z_{1\,O.C.}$ (2); b) when there is a short circuit at the output terminals of the TP - $Z_{1\,S.C.}$ (3):

$$Z_{1O.C.} = Z_{1-1'}\Big|_{Z_L=\infty} = Z_{input}\Big|_{Z_L=\infty} \quad (2)$$

$$Z_{1S.C.} = Z_{1-1'}\Big|_{Z_L=0} = Z_{input}\Big|_{Z_L=0} \quad (3)$$

The connections between the two impedances and the ABCD parameters of the A-matrix are as follows:

$$\begin{cases} A = C.Z_{1O.C.} \\ B = D.Z_{1S.C.} \end{cases} \quad (4)$$

After replacing (4) in (1) the following formulas for the A-parameters can extracted:

$$\begin{cases} A = \dfrac{Z_{1O.C.}(Z_{1S.C.}\dot{I}_1 - \dot{U}_1)}{(Z_{1S.C} - Z_{1O.C.})\dot{U}_2} \\ B = \dfrac{Z_{1S.C.}(\dot{U}_1 - Z_{1O.C.}\dot{I}_1)}{(Z_{1S.C} - Z_{1O.C.})\dot{I}_2} \\ C = \dfrac{Z_{1S.C.}\dot{I}_1 - \dot{U}_1}{(Z_{1S.C} - Z_{1O.C.})\dot{U}_2} \\ D = \dfrac{\dot{U}_1 - Z_{1O.C.}\dot{I}_1}{(Z_{1S.C} - Z_{1O.C.})\dot{I}_2} \end{cases} \quad (5)$$

The analysis of the explored ADAVR as a TP at the different levels of commutation was done by the help of the conditional linearization method.

The ADAVR (Fig. 1), considered as a TP with active load at forward transmission, was explored. The results about the values of the phasors of the input and the output currents and voltages for four intervals of the input supply voltage, which correspond to different closed switches ($S_1 \div S_4$) are presented in exponential form in Tables I ÷ IV.

TABLE I. INPUT AND OUTPUT CURRENTS AND VOLTAGES OF THE TP AT $U_1=E$=200,1V÷220V, R_L=3,4921Ω, S_1- CLOSED.

Parameter	Value	
E_1, V	220	200,1
$\dot{U}_1$, V	$220e^{j0}$	$200{,}1e^{j0}$
$\dot{I}_1$, A	$63{,}114\,e^{-j8{,}6681.10^{-1\circ}}$	$57{,}405\,e^{-j8{,}6681.10^{-1\circ}}$
$\dot{U}_2$, V	$220\,e^{-j3{,}782.10^{-15\circ}}$	$200{,}1\,e^{-j3{,}782.10^{-15\circ}}$
$\dot{I}_2$, A	$63\,e^{-j3{,}782.10^{-15\circ}}$	$57{,}301\,e^{-j3{,}782.10^{-15\circ}}$
$Z_{1\text{-}1'}$, Ω	$3{,}4857\,e^{j0{,}86681^\circ}$	

TABLE II. INPUT AND OUTPUT CURRENTS AND VOLTAGES OF THE TP AT $U1=E$=180,1V÷200V, R_L=3,4921Ω, S_2- CLOSED.

Parameter	Value	
E_1, V	200	180,1
$\dot{U}_1$, V	$200e^{j0}$	$180{,}1e^{j0}$
$\dot{I}_1$, A	$69{,}02e^{-j2{,}0437^\circ}$	$62{,}152e^{-j2{,}0437^\circ}$
$\dot{U}_2$, V	$219{,}1e^{-j1{,}5601^\circ}$	$197{,}3e^{-j1{,}5601^\circ}$
$\dot{I}_2$, A	$62{,}741e^{-j1{,}5601^\circ}$	$56{,}498e^{-j1{,}5601^\circ}$
$Z_{1\text{-}1'}$, Ω	$2{,}8977e^{j2{,}0437^\circ}$	

TABLE III. INPUT AND OUTPUT CURRENTS AND VOLTAGES OF THE TP AT $U_1=E$=160,1V÷180V, R_L=3,4921Ω, S_3- CLOSED.

Parameter	Value	
E_1, V	180	160,1
$\dot{U}_1$, V	$180e^{j0}$	$160{,}1e^{j0}$
$\dot{I}_1$, A	$74{,}851e^{-j9{,}875^\circ}$	$66{,}576e^{-j9{,}875^\circ}$
$\dot{U}_2$, V	$214{,}51e^{-j9{,}6522^\circ}$	$190{,}79e^{-j9{,}6522^\circ}$
$\dot{I}_2$, A	$61{,}428e^{-j9{,}6522^\circ}$	$54{,}637e^{-j9{,}6522^\circ}$
$Z_{1\text{-}1'}$, Ω	$2{,}4048e^{j9{,}875^\circ}$	

TABLE IV. INPUT AND OUTPUT CURRENTS AND VOLTAGES OF THE TP AT $U_1=E=150 \div 160V$, $R_L=3,4921\ \Omega$, S_4- CLOSED.

Parameter	Value	
E_1, V	160	150
$\dot{U}_1$, V	$160e^{j0}$	$150e^{j0}$
$\dot{I}_1$, A	$70,039\ e^{-j34,661^\circ}$	$65,662\ e^{-j34,661^\circ}$
$\dot{U}_2$, V	$178,24e^{-j34,57^\circ}$	$167,1\ e^{-j34,57^\circ}$
$\dot{I}_2$, A	$51,041\ e^{-j34,57^\circ}$	$47,851\ e^{-j34,57^\circ}$
$Z_{1-1'}$, Ω	$2,2844\ e^{j34,661^\circ}$	

The values of the ABCD parameters of the A-matrix were calculated on the basis of the received results for the input and the output currents and voltages of the TP for the different intervals of the input supply voltage, when only one of the switches $S_1 \div S_4$ is closed – Tables V ÷ VIII.

TABLE V. ABCD PARAMETERS OF THE A-SYSTEM AT $U_1=E=200,1V \div 220V$, $R_L=3,4921\Omega$, S_1 – CLOSED.

Coefficient	Algebraic form	Exponential form
A	$1+j1,2479.10^{-17}$	$1e^{j7,1498.10^{-16}\circ}$
B, Ω	$-1,8093.10^{-15}+j2,0275.10^{-16}$	$1,8206.10^{-15}\ e^{j173,61^\circ}$
C, S	$4,8751.10^{-4}-j4,3401.10^{-3}$	$4,3674.10^{-3}\ e^{-j83,591^\circ}$
D	$1+j2,4784.10^{-7}$	$1e^{j1,42.10^{-5}\circ}$

TABLE VI. ABCD PARAMETERS OF THE A-SYSTEM AT $U_1=E=180,1V \div 200V$, $R_L=3,4921\Omega$, S_2 – CLOSED.

Coefficient	Algebraic form	Exponential form
A	$9,1044.10^{-1}+j4.9275.10^{-5}$	$0,91044e^{j3,101.10^{-3}\circ}$
B, Ω	$7,2649.10^{-3}+j8,6619.10^{-2}$	$8,6923.10^{-2}\ e^{j85,206^\circ}$
C, S	$3,4882.10^{-4}-j2,4574.10^{-3}$	$2,482.10^{-3}\ e^{-j81,921^\circ}$
D	$1,0988-j7,0371.10^{-4}$	$1,0988e^{-j3,6694.10^{-2}\circ}$

TABLE VII. ABCD PARAMETERS OF THE A-SYSTEM AT $U_1=E=160{,}1V \div 180V$, $R_L=3{,}4921\Omega$, S_3 – CLOSED.

Coefficient	Algebraic form	Exponential form
A	$8{,}2219.10^{-1}+j1{,}3108.10^{-4}$	$0{,}82219e^{j9{,}1346.10^{-3}\circ}$
B, Ω	$1{,}7692.10^{-2}+j4{,}9086.10^{-1}$	$0{,}49118e^{j87{,}936^\circ}$
C, S	$4{,}0451.10^{-4}-j1{,}3629.10^{-3}$	$1{,}4216.10^{-3}e^{-j73{,}469^\circ}$
D	$1{,}2171+j1{,}1649.10^{-5}$	$1{,}2171e^{j5{,}4838.10^{-4}\circ}$

TABLE VIII. ABCD PARAMETERS OF THE A-SYSTEM AT $U_1=E=150 \div 160V$, $R_L=3{,}4921\Omega$, S_4 – CLOSED.

Coefficient	Algebraic form	Exponential form
A	$7{,}3038.10^{-1}+j1{,}5282.10^{-4}$	$0{,}73038e^{j1{,}1988.10^{-2}\circ}$
B, Ω	$3{,}075.10^{-2}+j1{,}7781$	$1{,}7783e^{j89{,}009^\circ}$
C, S	$3{,}4233.10^{-4}-j7{,}6102.10^{-4}$	$8{,}3447.10^{-4}e^{-j65{,}781^\circ}$
D	$1{,}371+j4{,}5923.10^{-4}$	$1{,}371e^{j1{,}9191.10^{-2}\circ}$

The A-matrix was used for the analysis of I_1, U_2 and I_2 at values of the load resistance R_L= 35,552Ω; 11Ω; 5,5Ω; 3,4921Ω for the interval of the input supply voltage $E=U_1$=220V÷160V and especially for E=220V, E=200V, E=180V, E=160V. The simulation results, which were received, coincide with those of the physical experiments with accuracy less than 1%. A comparison among the values of the input and the output quantities of the TP, obtained by simulations and by physical experiments at several values of the load resistance, is presented in Table IX.

TABLE IX. COMPARISON BETWEEN EXPERIMENTAL AND SIMULATION RESULTS

		Simulations			Experiments		
R_L, Ω	U_1, V	I_1, A	I_2, A	U_2, V	I_1, A	I_2, A	U_2, V
115,79	160	2,6724	1,8908	218,94	2,69	1,9	220
61,798	160	4,9306	3,5397	218,75	4,94	3,56	220
35,552	160	8,4945	6,1401	218,29	8,45	6,16	219

In the case, when the switch S_1 is closed, from the results obtained in Table V, it can be noted that the condition for reciprocity of the TP is fulfilled

($AD - BC = 1 + j2,4784.10^{-7} \approx 1$). Since $A \approx D$, it can be noted that the explored TP is symmetrical one.

In the cases of closed switches S_2, S_2 or S_4 the conditions for reciprocity ($AD - BC \approx 1$) are satisfied for the obtained A-matrices, but in these cases the TP is not symmetrical ($A \neq D$).

From the results, obtained for the A - matrices of the ADAVR considered as a TP, the following dependences are observed:

- The module of the parameter A decreases from 1 to 0,73038 and the phase of the same parameter A increases from $7,1498.10^{-16\circ}$ to $1,1988.10^{-2\circ}$;
- The module of the parameter B increases from $1,8206.10^{-15}\Omega$ to $1,7783\Omega$ and the phase of B decreases from 173,61° to about 85 ÷ 89°;
- The module of the parameter C decreases from 4,3674S to $8,3447.10^{-4}$S and the phase of C decreases from -83,591° to -65,781°;
- The module of the parameter D increases from 1 to 1,371, and the phase of D changes from $-3,6694.10^{-2\circ}$ to $1,9191.10^{-2\circ}$.

III. Conclusions

From the results of the explored ADAVR, considered as a TP, it can be summarized the following:

- At supply voltage in the range $E=U_1$=220V÷200,1V, the TP is symmetrical, and for the interval $E=U_1$=200V÷150V, the TP is not symmetrical.
- Due to the use of the conditional linearization method at the separate intervals of the input supply voltage, the reciprocity principle is valid for the obtained A-matrices, i.e. $AD - BC \approx 1$ despite the fact that the general model of the explored ADAVR is non-linear.
- The A-systems of equations of the explored ADAVR as a TP allow quick and accurate determination of the input or output quantities and parameters of such devices, which facilitate their practical application.
- The ABCD parameters of the explored ADAVR can be used not only in case when the load is active one but also when it is inductive or capacitive. That makes the proposed new approach more universal and useful.

Acknowledgment

The presented results are obtained by the support of the scientific project NP1/2019 of Scientific Research Fund at the Technical University of Varna, financed by the Ministry of Education and Science of Republic of Bulgaria.

References

[1] W. E. Kazibwe, and M. H. Senduala, Electric Power Quality Control Techniques, N. Y., Van Nostrand Reinhold, 1993.

[2] M. H. J. Bollen, Understanding Power Quality Problems: Voltage Sags and Interaptions, N. Y., IEEE Press, 2000.
[3] IEEE Std. C57.15-1999, IEEE Standard Requirements, Terminology and Test Code for Step-Voltage Regulators, 1999.
[4] J. H. Harlow, The Electric Power Engineering Handbook, Transformers, CRC Press LLC, 2001.
[5] S. Barudov, and E. Barudov, Discrete Alternating Current Regulators and Stabilizers, Sofia-Moscow, Pensoft, 2006. ISBN 978-954-642-276-7
[6] E. Barudov, Exploration and Analysis of the Electrical Processes in Circuits and Devices for Discrete Regulation of the Magnitude of AC Voltage, Auto-essay on PhD dissertation, Nikola Vaptsarov Naval Academy, Varna, 2014. (In Bulgarian)
[7] E. Barudov, S. Barudov, and E. Panov, "Discrete stabilizer of AC voltage", Patent BG1879 U1/12.05.2014. (in Bulgarian)
[8] E. Barudov, S. Barudov, and E. Panov, "Study of the electrical loading of semiconductor switching elements in a step voltage regulator", Acta Universitatis Pontica Euxinus, vol. 3, n. 1, 2004, pp. 97÷102. ISSN 1312-1669
[9] S. Barudov, and E. Panov, "Switching processes in a step voltage regulator", Acta Universitatis Pontica Euxinus, vol. 4, n. 1, 2005, pp. 21÷25. ISSN 1312-1669
[10] S. Barudov, E. Panov, and E. Barudov, "Parameters of the commutation process in a step-voltage AC regulator", Proceedings of Sankt Petersburg Institute for Qualification Promotion "Methods and Tools for Estimation of the State of Power Equipment", vol. 25, Sankt Petersburg, Russia, 2005, pp. 67-76. UDK 621.3.048, 621.315.62, BBK 31.264-04 (in Russian)
[11] E. Panov, E. Barudov, and M. Ivanova, "Study of the electrical characteristics of autotransformer discrete alternating voltage regulators with R-L loads, Proceedings of the XVI-th International Conference on Electrical Machines, Drives and Power Systems ELMA'2019, Varna, Bulgaria, 2019, pp. 321-325. ISBN 978-1-7281-1412-5, IEEE Catalog Number CFP19L07-USB
[12] E. Panov, E. Barudov, and M. Ivanova, "Exploration of the electric processes in discrete alternating step-voltage regulators", Proceedings of the XX-th International Symposium on Electrical Apparatus and Technologies SIELA'2018, Bourgas, Bulgaria, 2018, pp. 325 – 328. ISBN 978-1-5386-3418-9
[13] E. Barudov, E. Panov, and S. Barudov, "Analysis of electrical processes in alternating voltage control systems", Journal of International Scientific

Publication: Materials, Methods & Technologies, vol. 4, part 1, 2010, pp. 154÷182. ISSN 1313-2539

[14] E. Panov, E. Barudov, and S. Barudov, "Vector analysis and comparative valuation of precise and approximate non-linear models of discrete regulator with reducing input AC voltage". Proceedings of the XLVIII-th International Scientific Conference on Information, Communication and Energy Systems and Technologies ICEST'2013, vol. 1, Macedonia, 2013, pp. 771÷774. ISBN 978-9989-786-90-7

[15] N. Velikova, R. Dimitrova, M. Vasileva, and M. Yordanova, "Model research of the protection against arriving atmospheric surges in the substation 110 kV", Proceedings of the International Scientific Symposium "Electrical Power Engineering'2016", Varna, Bulgaria, pp. 78-82, 2016. ISBN (print) - 978-954-20-0760-9, ISBN (online) - 978-954-20-0762-3

[16] M. Mehmed-Hamza, M. Vasileva, A. Filipov, and M. Yordanova, "Some considerations for the choice of the work regime of the neutral of the power transformer of the electrical power networks medium voltage," Proceedings of the IV-th International Scientific Symposium "Electric Power Engineering'2007", 2007, pp. 169-171.

[17] M. Mehmed-Hamza, "Simulation models of the automatic transfer switch in the electrical power distribution network", Proceedings of the XLV-th International Scientific Conference on Information, Communication and Energy Systems and Technologies ICEST'2010, vol. 2, 2010, pp. 625–627.

[18] M. Simeonov, H. Ibrishimov, and P. Prodanov, "Modeling and analysis of an inductor – piece system with differentiated domains of the electromagnetic field in the inductor", PCIM Europe'2013, Paper № PP17, Nurnberg, Germany, 2013, pp. 1-5.

[19] D. Dankov, and P. Prodanov, "Analysis and design of quasi-resonant ZVS inverter for induction heating in a magnetic circuit", IEEE XXVI International Scientific Conference ELECTRONICS ET'2017, IEEE Conference Record #41615, Sozopol, Bulgaria, 2017, pp. 220-226.

Reference of Publications on the Topic

Publications in books, journals and conferences:

2003

1. Barudov E., Barudov S., Panov E., Switching Process in a Step Voltage Regulator, Annual Proceedings of Technical University in Varna, Publisher: High-Technology Park – TU Varna Ltd., (Proc. of Sc. Conf. of TU-Varna, Oct. 2003), 2003, pp. 5 – 10, ISSN 1312-1839.

2004

2. Barudov E., Barudov S., Panov E., Study of the Transient Process Length in Step Voltage Regulator, “Acta Universitatis Pontica Euxinus”, Vol. 3, N. 1, 2004, pp. 91÷96, ISSN 1312-1669.

3. Barudov E., Barudov S., Panov E., Study of the Electrical Loading of Semiconductor Switching Elements in a Step Voltage Regulator, “Acta Universitatis Pontica Euxinus”, Vol. 3, N. 1, 2004, pp. 97÷102, ISSN 1312-1669.

4. Barudov S., Panov E., Study of the loading of the switching elements in a step voltage regulator, Annual Proceedings of Technical University of Varna, Varna, 2004, pp. 117÷122.

5. Barudov S., Panov E., Influence of load character on duration of switching processes in step regulator of alternating voltage, Scientific Conference "Information, Innovations, Investments", Perm, Russia, 2004, pp. 134÷138, ISBN 5-93978-024-5. (in Russian)

6. Barudov S., Panov E., About some parameters of switching process in step regulator of alternating voltage, Scientific Conference "Information, Innovations, Investments", Perm, Russia, 2004, pp. 138÷143, ISBN 5-93978-024-5. (in Russian)

7. Barudov S., Panov E., Study of the Switching Processes in a Step Voltage Regulator, Annual Proceedings of Technical University of Varna, 2004, pp. 111÷116, ISSN 1312-1839.

8. Barudov S., Panov E., Study of the Loading of the Switching Elements in a Step Voltage Regulator, Annual Proceedings of Technical University of Varna, 2004, pp. 117÷122, ISSN 1312-1839.

2005

9. Barudov S., Panov E., Switching Processes in a Step Voltage Regulator, "Acta Universitatis Pontica Euxinus", Vol. 4, N. 1, 2005, pp. 21÷25, ISSN 1312-1669.

10. S. Barudov, E. Panov, and E. Barudov, Parameters of the commutation process in a step-voltage AC regulator, Proceedings of Sankt Petersburg Institute for Qualification Promotion SPb: PEIPK "Methods and Tools for Estimation of the State of Power Equipment", Sankt Petersburg, Russia, Vol. 25, 2005, pp. 67-76, UDK 621.3.048, 621.315.62, BBK 31.264-04. (in Russian)

2007

11. Barudov E., Panov E., Barudov S., Exploration of Precise Non-Linear Model of Discrete Autotransformer Step-Voltage AC Regulator with Semiconductor Commutators, Annual of TU-Varna, 2007, Bulgaria, ISSN: 1311-896X, pp. 3÷9. (in Bulgarian)

12. Barudov E., Panov E., Barudov S., Study of a precise non-linear model of an autotransformer discrete voltage regulator with semiconductor switching elements, Annual of TU-Varna, 2007, pp. 91÷96.

13. Barudov E., Panov E., Barudov S., Study of a precise non-linear model of an autotransformer discrete voltage regulator with semiconductor switching elements, Annual of TU-Varna, 2007, pp. 3÷9, ISSN 1311-896X. (in Bulgarian)

2009

14. Panov E., Barudov E., Barudov S., Advanced algorithm for the analysis of a precise non-linear model of an autotransformer discrete voltage regulator with semiconductor switching elements, Annual of TU-Varna, Vol. I, 2009, pp. 47-52, ISSN: 1311-896X. (in Bulgarian)

15. Panov E., Barudov E., Barudov St., Automated computer program AVTO for simulation of processes in autotransformer discrete variable voltage step-up regulators, Annual of the Technical University - Varna, Vol. I, 2009, pp. 53 – 58, ISSN: 1311-896X. (in Bulgarian)

2010

16. Barudov E., Panov E., Barudov S., Analysis of Electrical Processes in Alternating Voltage Control Systems, 12-th International Symposium "Materials, Methods&Technologies (MMT), 11-15 June 2010, Sunny Beach, Bulgaria, published in Journal of International Scientific Publications: Materials, Methods

& Technologies, Volume 4, Part 1, 2010, pp. 154 – 182, ISSN: 1313-2539, http://www.science-journals.eu.

17. Barudov S., Panov E., Barudov E., Analysis of electrical processes in a discrete AC voltage regulator under resistive-capacitive load, Proceedings of the International Scientific and Technical Conference "Electrical Power Engineering 2010", 14-16 October 2010, Varna, pp. 332-341, ISBN 978-954-20-0497-4. (in Bulgarian)

18. Barudov E., Panov E., Barudov S., Analysis of electrical processes in a discrete AC voltage regulator with active-inductive load, Annual of the Technical University - Varna, Vol. II, 2010, pp. 30 - 35, ISSN: 1311-896X. (in Bulgarian)

2011

19. Barudov E., Panov E., Barudov S., Comparative study of the switching processes in autotransformer discrete step-up voltage regulator. ELMA 2011, pp. 55÷60, ISSN 1313-4965. (in Bulgarian)

20. Barudov E., Panov E., Barudov S., Commutation Processes at Reactive Loads in a Discrete Alternating Voltage Regulator, Annual of the Technical University of Varna, Vol. I, 2011, pp. 3- 8, ISSN: 1311-896X. (in Bulgarian)

2013

21. Panov E., Barudov E., Barudov S., Vector Analysis and Comparative Valuation of Precise and Approximate Non-Linear Models of Discrete Regulator with Reducing Input AC Voltage, Proceedings of XLVIII-th International Scientific Conference on Information, Communication and Energy Systems and Technologies ICEST'2013, 26-29 June 2013, Ohrid, Macedonia, Vol. 2, 2013, pp. 771-774, ISBN 978-9989-786-89-1.

22. Barudov S., Barudov E., Panov E., Study of the peculiarities of the commutation processes in a discrete altenating voltage regulator at asynchronous switching mode, Marine Scientific Forum, Electronics, Electrical Engineering and Automatics. Informatics, N. Y. Vaptsarov", Vol. 4, 2013, pp. 104-109, ISSN 1310-9278. (in Bulgarian)

23. Barudov S., Panov E., Barudov E., Study of the sensitivity of the models of autotransformer discrete alternating voltage regulator, Marine Scientific Forum, Electronics, Electrical Engineering and Automatics. Informatics, VVMU "N. Y. Vaptsarov", Vol. 4, 2013, pp. 97-103, ISSN 1310-9278. (in Bulgarian)

24. Panov E., Barudov E., Barudov S., Error analysis of the vector analysis and vector measurements in an autotransformer discrete variable voltage regulator,

Jubilee Scientific Conference "50 Years of ETET Department", 4-5 October 2013, Varna, Bulgaria, Annual of the Technical University - Varna, Vol. I, 2013, pp. 180-184, ISSN: 1311-896X. (in Bulgarian)

25. Barudov S., Panov E., Barudov E., Ivanova M., Discrete stabilizer of AC voltage, Patent BG1727 U1 /07.08.2013, (utility model patent). (in Bulgarian)

2014

26. Barudov S., Panov E., Barudov E., Ivanova M., Discrete stabilizer of AC voltage, Patent, BG1879 U1 /12.05.2014, (utility model patent). (in Bulgarian)

2018

27. Panov E., Barudov E., Ivanova M., Exploration of the Electric Processes in Discrete Alternating Step-Voltage Regulators, Proceedings of the XX-th International Symposium on Electrical Apparatus and Technologies SIELA'2018, Bourgas, Bulgaria, 3-6 June 2018, pp. 325 - 328, ISBN 978-1-5386-3418-9.

28. Panov E., Ivanova M., Barudov E., Study of the Electrical Characteristics of Autotransformer Discrete Alternating Voltage Regulators with R-L Loads, Proceedings of the XVI-th International Conference on Electrical Machines, Drives and Power Systems ELMA'2019, 6-8 June 2019, Varna, Bulgaria, pp. 321-325, ISBN 978-1-7281-1412-5.

29. Panov E., Barudov E., Ivanova M., Exploration of the Two-Port Parameters of Autotransformer Discrete Alternating Voltage Regulators, Proceedings of the XI-th Electrical Engineering Faculty Conference BulEF'2019, 11-14 September 2019, 4 pages, Electronic ISBN: 978-1-7281-2697-5, DOI: 10.1109/BulEF48056.2019.9030701.

2020

30. Panov E., Dimitrov D., Barudov E., Ivanova M., Protection Regimes and Electrical Quantities in Autotransformer Alternating Voltage Regulators, Proceedings of the XXI-th International Symposium on Electrical Aparatus and Technologies SIELA'2020, 3-6 June 2020, Bourgas, Bulgaria, pp. 338-341, Electronic ISBN: 978-1-7281-4346-0, USB ISBN: 978-1-7281-4345-3, Print on Demand (PoD) ISBN: 978-1-7281-4347-7, DOI: 10.1109/SIELA49118.2020.9167144.

31. Ivanova M., Panov E., Barudov E., Exploration of the Efficiency of Autotransformer Discrete Alternating Voltage Regulators, IEEE International Conference Automatics and Informatics ICAI'2020, 1-3 October 2020, Varna, Bulgaria, 5 pages. SBN 978-1-7281-9308-3, DOI: 10.1109/ICAI50593.2020.9311348.

32. Barudov E., Panov E., Comparative Valuation of Precise and Approximate Non-Linear Models of an AC Discrete Voltage Regulator and Vector Analysis of Its Parameters, South Florida Journal of Development, Miami, USA, oct./dec. 2020, vol. 1, n. 4, pp. 202-210, ISSN 2675-5459, DOI: 10.46932/sfjdv1n4-003.

2021

33. Panov E., Ivanova M., Barudov E., Study of the Parameters of an Autotransformer Discrete Alternating Voltage Regulators Considered as a Two-Port, Proceedings of the 17-th International Conference on Electrical Machines, Drives and Power Systems (ELMA'2021), 1-4 July 2021, Sofia, Bulgaria, 4 pages, Electronic ISBN: 978-1-6654-3582-6, USB ISBN: 978-1-6654-3581-9, Print on Demand (PoD) ISBN: 978-1-6654-1186-8, INSPEC Accession Number: 21015774, DOI: 10.1109/ELMA52514.2021.9503020.

2022

34. Barudov E., Panov E., Comparative Valuation of Precise and Approximate Non-Linear Models of an AC Discrete Voltage Regulator and Vector Analysis of Its Parameters, in the e-book "Technologies impacts in exact sciences", edited by Catapan Edilson Antonio, I-st edition, South Florida Publishing - Miami, USA, 2022, pp. 21-31, ISBN: 978-1-7361138-6-8, DOI: 10.47172/sfp2020.ed.0000028, (Reprint from South Florida Journal of Development, Miami, USA, oct./dec. 2020, vol. 1, n. 4, pp. 202-210, ISSN 2675-5459, DOI: 10.46932/sfjdv1n4-003).

35. Barudov E., Panov E., Ivanova M., Research into the Z-Parameters and the Characteristic Parameters of a Ship's Alternating Voltage Regulators, Proceedings of the XXII-th International Symposium on Electrical Apparatus and Technologies (SIELA'2022), Publisher IEEE, Bourgas, Bulgaria, 1-4 June 2022, pp. 9 – 12, INSPEC Accession Number: 21974600, ISBN: 978-1-6654-1139-4/22, DOI: 10.1109/SIELA54794.2022.9845782.

36. Panov E., Barudov E., Ivanova M., Two-port parameters of autotransformer discrete alternating voltage regulators, South Florida Journal of Development, Miami, USA, v. 3, n. 6. pp. 7049-7063, nov/dec., 2022, ISSN 2675-5459, DOI: 10.46932/sfjdv3n6-051.

2023

37. Panov E., Barudov E., Ivanova M., Two-port parameters of autotransformer discrete alternating voltage regulators, Chapter 8 in the e-book: "Methodology Focused on the Area of Interdisciplinarity", 16 pages, Published on 08.02.2023, ISBN: 978-65-84976-19-1, Seven Publicações LTDA, Brazil, CNPJ:

43.789.355/0001-14, DOI: https://doi.org/10.56238/methofocusinterv1-008 (Reprint from South Florida Journal of Development, Miami, USA, v. 3, n. 6. pp. 7049-7063, nov/dec., 2022. ISSN 2675-5459, DOI: 10.46932/sfjdv3n6-051).

www.ingramcontent.com/pod-product-compliance
Lightning Source LLC
Chambersburg PA
CBHW070723220326
41598CB00024BA/3280
9798886769463